This Book Comes With Lots of
FREE Online Resources

Nolo's award-winning website has a page dedicated just to this book. Here you can:

DOWNLOAD FORMS – Access forms and worksheets from the book online

KEEP UP TO DATE – When there are important changes to the information in this book, we'll post updates

GET DISCOUNTS ON NOLO PRODUCTS – Get discounts on hundreds of books, forms, and software

READ BLOGS – Get the latest info from Nolo authors' blogs

LISTEN TO PODCASTS – Listen to authors discuss timely issues on topics that interest you

WATCH VIDEOS – Get a quick introduction to a legal topic with our short videos

And that's not all.
Nolo.com contains thousands of articles on everyday legal and business issues, plus a plain-English law dictionary, all written by Nolo experts and available for free. You'll also find more useful **books, software, online apps, downloadable forms,** plus a **lawyer directory.**

With
Downloadable
FORMS

NOLO
LAW for ALL

Get forms and more at
www.nolo.com/back-of-book/HICI.html

FEB 15

8th Edition

Working With Independent Contractors

Stephen Fishman, J.D.

NOLO
LAW for ALL

EIGHTH EDITION   JULY 2014

Editor   MICAH SCHWARTZBACH

Cover Design   SUSAN PUTNEY

Production   COLLEEN CAIN

Proofreading   SUSAN CARLSON GREENE

Index   ELLEN SHERRON

Printing   BANG PRINTING

Fishman, Stephen, author.
 Working with independent contractors / Stephen Fishman. — Eighth edition.
     pages cm
 Summary: "With Working With Independent Contractors, you can learn how to: create a
valid contract, assess who qualifies as an independent contractor, and handle an IRS audit"—
Provided by publisher.
 ISBN 978-1-4133-2046-6 (paperback) — ISBN 978-1-4133-2047-3 (epub ebook)
 1. Independent contractors—Legal status, laws, etc.—United States—Popular works.
 2. Contracts for work and labor—United States—Popular works. 3. Independent
contractors—Legal status, laws, etc.—United States—Forms. 4. Contracts for work and
labor—United States—Forms. I. Title.
 KF898.F57 2014
 346.7302'4—dc23
                         2014004375

This book covers only United States law, unless it specifically states otherwise.

Please note

We believe accurate, plain-English legal information should help you solve many of
your own legal problems. But this text is not a substitute for personalized advice
from a knowledgeable lawyer. If you want the help of a trained professional—and
we'll always point out situations in which we think that's a good idea—consult an
attorney licensed to practice in your state.

Acknowledgments

Many thanks to:

Barbara Kate Repa, Amy DelPo, Marguerite Fa-Kaji, JinAh Lee, and Micah Schwartzbach for their superb editing.

Jake Warner for his editorial contributions.

Fred Daily for many helpful comments on federal tax law.

Ellen Sherron for the helpful index.

Susan Carlson Greene for thorough proofreading.

About the Author

Stephen Fishman has dedicated his career as an attorney and author to
writing useful, reliable, and recognized guides on taxes and business law
for small businesses, entrepreneurs, independent contractors, and the
self-employed. He is the author of 20 books, including the following, all
published by Nolo:

- *Working for Yourself: Law & Taxes for Independent Contractors,
  Freelancers & Consultants*
- *Consultant & Independent Contractor Agreements*
- *Deduct It! Lower Your Small Business Taxes*
- *Home Business Tax Deductions: Keep What You Earn*, and
- *Tax Deductions for Professionals*.

He has also written hundreds of articles and been quoted in *The New
York Times*, *The Wall Street Journal*, *Chicago Tribune*, and many other
publications. His website is www.fishmanlawandtaxfiles.com.

Table of Contents

Your Legal Companion for Working With Independent Contractors......1

1 Benefits and Risks of Working With Independent Contractors......3
Benefits of Using Independent Contractors......4
Risks of Using Independent Contractors......8

2 The Common Law Test......15
When a Legal Test Is Necessary......17
The Right of Control Is Key......19
Factors for Measuring Control......20
Evaluating a Worker's Status......33

3 How the IRS Classifies Workers......39
Four Steps to Classification Under the IRS Rules......40
Step 1: Check Statutory Independent Contractor Rules......46
Step 2: Analyze the Worker Under the Common Law Test......49
Step 3: Check Statutory Employee Rules......72
Step 4: Check the Safe Harbor Rules......80

4 IRS Audits......99
Why Audits Occur......101
Audit Basics......102
The Classification Settlement Program......109
Voluntary Classification Settlement Program......113
IRS Assessments for Worker Misclassification......116
Penalties for Worker Misclassification......122
Interest Assessments......125

Criminal Sanctions...125

Obamacare Penalties..126

Retirement Plan Audits..126

Worker Lawsuits for Pensions and Other Benefits.......................................128

5 **State Payroll Taxes**..129

State Unemployment Compensation...130

State UC Classification Tests..134

State Disability Insurance 150

State Income Taxes...153

6 **Workers' Compensation**..155

Basics of the Workers' Compensation System 156

Who Must Be Covered...160

Exclusions From Coverage..160

Classifying Workers for Workers' Compensation Purposes.......................166

If Your Workers Are ICs..178

Obtaining Coverage...181

7 **Hiring Household Workers and Family Members**...................................183

Household Workers...184

Family Members as Workers...198

8 **Health, Safety, Labor, and Antidiscrimination Laws**..........................203

Obamacare (Affordable Care Act)...206

Federal Wage and Hour Laws..215

Federal Labor Relations Laws..222

Family and Medical Leave Act 223

Fair Credit Reporting Act..224

Antidiscrimination Laws 225

Worker Safety Laws..229

Immigration Laws...230

9   Intellectual Property Ownership 231

What Is Intellectual Property? 232

Laws Protecting Intellectual Property 232

Copyright Ownership 234

Trade Secret and Patent Ownership 241

10   Strategies for Avoiding Trouble 243

Hiring Incorporated Independent Contractors 244

Employee Leasing 252

11   Procedures for Working With Independent Contractors 261

Before Hiring an IC 262

While the IC Works for You 267

After the IC's Services End 271

12   Independent Contractor Agreements 281

Using Written Agreements 283

Drafting Agreements 285

Essential Provisions 292

Optional Provisions 318

Sample IC Agreement 325

13   Help Beyond This Book 333

Finding and Using a Lawyer 334

Help From Other Experts 336

Doing Your Own Legal Research 337

Appendixes

A   How to Use the Interactive Forms 345

Editing RTFs 346

List of Forms 347

B **Contractor Screening Documents** 349

   Independent Contractor Questionnaire 351

   Documentation Checklist 353

   Worker Classification Checklist 355

Index 359

Your Legal Companion for Working With Independent Contractors

Many people hire independent contractors for help at home or at work. Independent contractors are not your employees—they have their own businesses, usually have their own tools, often work "by the job" instead of by the hour, and work for many people besides you. They cover a lot of ground, including accountants or bookkeepers who help prepare your taxes, janitors or housekeepers who clean your office or residence, building contractors who remodel your house, and computer consultants who design programs for use by your company. According to the Bureau of Labor Statistics, more than 10 million American workers are independent contractors (ICs), comprising 7.4% of the entire workforce. ICs do every conceivable type of work: 20.5% are in executive, administrative, and managerial positions; 18.9% perform precision production, craft, and repair jobs; 18.5% do professional specialty jobs; and 17.3% are in sales.

Using an independent contractor instead of hiring an employee gives you flexibility (the IC works with you periodically or only once, instead of all the time) and is financially attractive. With employees, you have to withhold and pay state and federal income taxes, along with Social Security and Medicare taxes, and you need to pay for unemployment compensation, workers' compensation, and state disability insurance. ICs don't come with these tasks and costs. But here's the rub: As you try to save money, you can't simply designate a worker as an IC. A worker's status as an IC or an employee depends not on what you call the person, but on how the government—the IRS or state taxing and other agencies—views his or her work. And the government clearly has an incentive to classify more workers as employees than as independent contractors: The more taxes an employer pays, the more funds the government receives.

This doesn't mean that you can't hire independent contractors; it just means that you should know the guidelines the government uses to classify workers, so that you can prove that the independent contractors you hire shouldn't be considered your employees. This book provides you with all of the information you need, including:

- how to determine whether a worker is an independent contractor or an employee, by satisfying any and all of the tests developed by the IRS and state agencies
- how to document that a worker is truly an independent contractor
- how to draft the written agreement between yourself and the independent contractor, and
- how to reduce your chances of facing, or losing, an audit by the IRS or another government agency.

Staying within the good graces of the IRS and other government agencies isn't your only goal when you hire independent contractors. You'll be well served by this book's quick review of discrimination law, which will help you avoid the possibility that your IC will accuse you of workplace discrimination. And, if the IC creates "intellectual property" for you—an instruction manual, a piece of artwork, or advertising copy, for example—you'll learn in these pages what steps to take to become its owner.

Hiring independent contractors can provide many benefits to you, both personally and professionally. Putting this book's lessons into practice, you can pursue such working relationships with confidence, knowing that your rights are properly protected.

Benefits and Risks of Working With Independent Contractors

Benefits of Using Independent Contractors 4

    Financial Savings.. 4

    Reduced Exposure to Lawsuits... 6

    Flexibility in Hiring.. 7

Risks of Using Independent Contractors.. 8

    Federal Audits.. 8

    State Audits.. 10

    Worker Misclassification Lawsuits.. 11

    Loss of Control 11

    Loss of Continuity 11

    Restrictions on Right to Fire 12

    Liability for Injuries 12

    Possible Loss of Copyright Ownership.. 12

There are many benefits to hiring ICs, but there are serious risks as well. No book can tell you whether you should use ICs in your business, but this chapter will help you make an informed decision by summarizing the potential advantages and disadvantages.

Benefits of Using Independent Contractors

It can cost less to use ICs instead of employees because you don't have to pay employment taxes and various other employee expenses for ICs. In addition, you will be less vulnerable to some kinds of lawsuits. Perhaps most importantly, hiring ICs gives you greater flexibility to expand and contract your workforce as needed.

Financial Savings

It usually costs more to hire employees than ICs because, in addition to employee salaries or other compensation, you will have to pay a number of employee expenses. On average, these expenses add 30% to your payroll costs. For example, if you pay an employee $10 per hour, you must pay an additional $3 per hour in employee expenses.

You incur none of these expenses when you hire an IC. Even though ICs are often paid more per hour than employees doing the same work, you will still save money in the long run by using the former.

In addition to the costs of payroll processing, the most common employee expenses include:

- federal payroll taxes
- unemployment compensation insurance
- workers' compensation insurance
- office space and equipment, and
- employee benefits like paid vacation and health insurance.

Federal Payroll Taxes

Employers must withhold and pay federal payroll taxes for employees. They must pay a 7.65% Social Security and Medicare tax and a small—usually 0.8%—federal unemployment tax out of their own pockets. In addition, employers must withhold Social Security and Medicare taxes and federal income taxes from their employees' paychecks, and periodically hand this money over to the IRS. (See Chapter 3.)

In contrast, you don't have to withhold or pay any federal payroll taxes for ICs. This will help you save money, not only in taxes, but in bookkeeping costs as well.

Unemployment Compensation

Employers in every state are required to contribute to a state unemployment insurance fund on behalf of most employees. The unemployment tax rate is usually somewhere between 2% and 5% of employee wages, up to a maximum amount set by state law. (See Chapter 5 for more on unemployment compensation rules.)

Workers' Compensation Insurance

Employers must provide workers' compensation insurance coverage for most types of employees in order to provide some wage replacement and reimbursement of medical bills if an employee is injured on the job. Depending on the state, employers can get workers' compensation insurance either from private insurers or state workers' compensation funds. Premiums can range from a few hundred dollars per year to thousands, depending upon the employee's occupation and the company's claims history. Employers don't have to carry workers' compensation insurance for ICs. (See Chapter 6 for information about state workers' compensation laws.)

Office Space and Equipment

Employers typically provide their employees with workspace and whatever equipment they need to do their jobs. This isn't necessary for ICs, who ordinarily provide their own workplaces and equipment. Office space is usually an employer's second biggest expense; only employee salaries and benefits cost more.

Obamacare

Starting in 2015, the Affordable Care Act (also known as "Obamacare") requires that all employers with the equivalent of more than 100 full-time employees (50 in 2016 and later) provide them with minimally adequate health insurance or pay a penalty to the IRS. The Obamacare employer mandate does not apply to independent contractors—they don't count toward the applicable employee threshold. Thus, using ICs can save a hiring firm substantial sums on health insurance. (See Chapter 8 for detailed discussion of Obamacare.)

Employee Benefits Other Than Health Care Insurance

Employers usually provide their employees with benefits such as paid vacation, sick leave, retirement benefits, and life or disability insurance. You need not—and should not—provide ICs with such benefits.

Reduced Exposure to Lawsuits

When you hire employees, you may be subject to some types of legal claims that ICs can't make against you.

Labor and Antidiscrimination Laws

Employees have a wide array of rights under state and federal labor and antidiscrimination laws. Among other things, these laws:

- impose a minimum wage and require many employees to be paid time-and-a-half for overtime
- make it illegal for employers to discriminate against employees on the basis of race, religion, gender, national origin, age, or disability
- protect employees who wish to unionize, and
- make it unlawful for employers to knowingly hire illegal aliens.

In recent years, a growing number of employees have brought lawsuits against employers alleging violations of these laws. Some employers have had to pay hefty damages to their employees. In addition, various watchdog agencies, such as the U.S. Department of Labor and the U.S. Equal Employment Opportunity Commission, have authority to take administrative or court action against employers who violate these laws.

Few of these antidiscrimination and employment laws apply to ICs, so you have much less exposure to these kinds of employee claims and lawsuits when you use ICs instead of employees. (See Chapter 8.)

Wrongful Termination Liability

Employees can also sue for wrongful termination. In these legal actions, an employee claims that his or her firing was illegal or constituted a breach of contract. Wrongful termination laws vary from state to state. Under some circumstances, for example, it might be a breach of contract for you to fire an employee without good cause. To guard against wrongful termination claims, employers must carefully document the reasons for firing an employee so they can defend their actions in court, if necessary.

ICs cannot bring wrongful termination lawsuits. However, there usually are contractual restrictions on when you can fire an IC. For example, your contract might state that you can fire an IC only with written notice, or only for his or her failing to meet obligations under the contract. If you disregard these limits, you could face a breach-of-contract lawsuit.

Liability for Workers' Actions

When you hire an employee, you're liable for anything he or she does within the scope of employment. For example, if an employee gets into an auto accident while making a delivery for work, you may be liable for the damages.

Subject to several important exceptions, this is not the case with ICs. You are not liable for an IC's actions, work-related or not, unless one of the following is true:

- The IC you hired was not qualified to do the job and you were negligent in hiring him or her.
- An injury occurs because of your improper instructions to the IC.
- You know the IC is violating the law in working for you—for example, you hire an unlicensed IC to perform work that requires a construction contractor license.
- You hire an IC to do work that is inherently dangerous—for example, building demolition.

Flexibility in Hiring

Working with ICs provides a level of flexibility that you just can't get from employees. You can hire an IC to accomplish a specific task, which gives you specialized expertise for a short period. You need not go through the trauma and potential severance costs (and lawsuits) of having to lay off or fire an employee. And an experienced IC can usually be productive immediately, eliminating the time and expense of training. By using ICs, you can expand and contract your workforce as needed, quickly and inexpensively.

Risks of Using Independent Contractors

After reading about the possible benefits of using ICs, you might be thinking: "I'll never hire an employee again; I'll just use independent contractors." But be aware that there are some substantial risks involved in classifying workers as ICs.

Federal Audits

The IRS wants to see as many workers as possible classified as employees, not ICs, so that it can immediately collect taxes based on payroll withholding. Also, the IRS believes that ICs are more likely to underreport their income when tax time rolls around. In recent years, the IRS has mounted an aggressive attack on employers who, in its view, misclassify employees as ICs.

If the IRS audits your business and determines that you have misclassified employees as ICs, it may impose substantial interest and penalties. Such assessments can easily put a small company out of business. The owners of an unincorporated business may be held personally liable for such assessments and penalties. Even if your business is a corporation, you could still be held personally liable for the tax, interest, and penalties.

Other agencies can also audit businesses for misclassifying employees. These include the Department of Labor, which enforces the federal minimum wage and hours laws; the National Labor Relations Board, which enforces employees' federal right to unionize; and the Occupational Safety and Health Administration, which enforces workplace safety laws.

The Department of Labor has taken particular interest in the worker misclassification issue. It recently hired hundreds of investigators and other enforcement staff to undertake a "Misclassification Initiative" designed to recover back wages and other benefits from employers who misclassified their workers as ICs.

(See Chapter 8 for information about labor and antidiscrimination laws.)

The following chart, prepared by the General Accounting Office, shows the principal government agencies that are concerned with misclassification of employees as ICs.

Legal Implications of Employment Classification		
Entity	**Law**	**Areas Potentially Affected by Employee Misclassification**
U.S. Department of Labor	Fair Labor Standards Act	Minimum wage, overtime, and child labor provisions
	Family and Medical Leave Act	Job-protected and unpaid leave
	Occupational Safety and Health Act	Safety and health protections
U.S. Department of Treasury— Internal Revenue Service	Federal tax law, including: Federal Insurance Contributions Act, Federal Unemployment Tax Act, Self-Employment Contributions Act	Federal income and employment rates
U.S. Department of Health and Human Services	Title XVIII of the Social Security Act (Medicare)	Medicare benefit payments
DOL/IRS/Pension Benefit Guaranty Corporation	Employee Retirement Income Security Act	Pension, health, and other employee benefit plans
Equal Employment Opportunity Commission	Title VII of the Civil Rights Act	Prohibitions of employment discrimination based on race, color, religion, gender, or national origin
	Americans with Disabilities Act	Prohibitions of discrimination against individuals with disabilities
	Age Discrimination in Employment Act	Prohibitions of employment discrimination against any individual 40 years of age or older
National Labor Relations Board	National Labor Relations Act	The right to organize and bargain collectively
Social Security Administration	Social Security Act	Retirement and disability payments

Legal Implications of Employment Classification (continued)		
DOL/State agencies	Unemployment Insurance law	Unemployment Insurance benefit payments
State agencies	State tax law	State income and employment taxes
	State workers' compensation law	Workers' compensation benefit payments

State Audits

Audits by state agencies are even more common than federal audits. State audits most frequently occur when workers who were classified as ICs apply for unemployment compensation after their services are terminated. Your state unemployment compensation agency will begin an investigation, and you may be subject to fines and penalties if it determines that your workers should have been classified as employees for unemployment compensation purposes.

If workers classified as ICs are injured on the job and apply for workers' compensation benefits, you can expect an audit by your state workers' compensation agency. Very substantial fines and penalties can be imposed on businesses that misclassify employees as ICs for workers' compensation purposes. You may even face a court order preventing you from doing business until you obtain workers' compensation insurance. (See Chapter 6 for more about state workers' compensation laws.)

Moreover, several states, including Colorado, Delaware, Illinois, Indiana, Maryland, Minnesota, New Hampshire, New Jersey, Rhode Island, and Washington, have adopted laws that make it "fraud" for an employer to either "knowingly" misclassify its workers as ICs to avoid providing them with unemployment or workers' compensation insurance, or fail to comply with federal or state prevailing wage and overtime pay rules.

Although not as common as unemployment insurance or workers' compensation audits, your state tax agency may also audit to ensure that your workers are properly classified for purposes of your state income tax law. Again, fines and penalties may be imposed for misclassifying employees as ICs. (See Chapter 5 for more information about state tax laws.)

Worker Misclassification Lawsuits

A growing number of worker misclassification lawsuits are being filed by workers under state and federal employment laws. Many of these take the form of class action suits in which plaintiffs' lawyers represent tens, hundreds, or even thousands of similarly situated workers.

The plaintiffs in these cases seek payment for employee benefits such as minimum wages, overtime pay, sick leave, health care, and vacation pay. Such lawsuits have been filed in recent years on behalf of a wide variety of workers, including truckers, delivery workers, insurance agents, janitors, telecommunications support personnel, newspaper delivery carriers, health care professionals, "crowdsourced" workers, and even exotic dancers.

Loss of Control

Another possible drawback to classifying workers as ICs is that you lose control over them. Unlike employees, whom you can closely supervise and micromanage, you have to leave independent contractors alone to do the job you are paying them to do. If you help them too much or interfere too much in their performance, you risk turning them into employees. (See Chapter 2 for more about this control issue.)

Some business owners want to be in charge of everything and everybody involved with their business. If you're one of them, and you want to control how your workers do their jobs, classify them as employees.

Loss of Continuity

Generally, employers use particular ICs only as needed for short-term projects. This can result in workers constantly coming and going, which can be inconvenient and disruptive for any workplace. And the quality of work you get from various ICs may be uneven. One reason businesses hire employees is to be able to depend on the same workers day after day.

Restrictions on Right to Fire

You do not have an unrestricted right to fire an IC as you do with most employees. Your right to terminate an IC's services is limited by the terms of your agreement. If you terminate an IC who performs adequately and otherwise satisfies the terms of the agreement, you'll be liable to him or her for breaking the agreement. In other words, the IC can sue you and get an order requiring you to pay a substantial amount of money in damages.

Liability for Injuries

Employees covered by workers' compensation who are injured on the job cannot sue you for damages. Instead, they can file workers' compensation claims and receive workers' compensation benefits. This is not the case with ICs. They can sue you for damages if they claim they were injured because of your negligence, such as your failure to provide a safe workplace. If the injuries are substantial and your negligence is clear, you may end up having to pay quite a bit of money in damages. When you hire ICs who perform services at your place of business, you should have liability insurance to cover the costs of such lawsuits. Depending on your situation, this may or may not be cheaper than obtaining workers' compensation insurance.

Possible Loss of Copyright Ownership

If you hire ICs to create works that can be copyrighted—for example, book chapters or photographs—you will not own the legal rights to the work unless you use written agreements transferring copyright ownership to you in advance. This is not the case with employees. (See Chapter 9 for information about intellectual property issues.)

Ten Myths About Hiring Independent Contractors

Common misconceptions about classifying workers include the following:

1. If I issue IRS Form 1099-MISC, the worker is an IC.
An IRS Form 1099-MISC is simply a method the government uses to track and report certain types of nonemployment income. When you provide an IRS Form 1099-MISC to a worker for payment of services, it does not automatically make the worker an IC.

2. Any worker I pay less than $600 in a year is an IC.
The amount paid to a worker is not, by itself, a factor in determining whether he or she is an employee or IC.

3. Part-time and temporary workers are always ICs.
Employees can and do work part-time and short-term.

4. A signed contractor agreement makes a worker an IC.
A signed IC agreement can help, but it will never by itself make a worker an IC. The actual practices of the hiring firm and worker are more important than the wording of an agreement.

5. Everyone else is doing it, so it's okay for me to treat my workers as ICs.
Other drivers often speed, but that isn't a defense if you get a ticket.

6. Workers who perform similar work for other businesses are always ICs.
The relationship each worker has with each business is evaluated independently. The same worker can be an IC for one business and an employee for another.

7. A worker who has a business license and business card is an IC.
They can help, but a business license and a business card, by themselves, do not make a worker an IC. All the circumstances need to be considered.

8. Workers paid by commission are always ICs.
The method of payment is not, by itself, a determining factor. Employees are often paid by commission, too.

9. Workers who perform offsite are always ICs.
Many employees work at home, at least part of the time. With today's technological capabilities, off-site work is consistent with any type of worker.

10. You're safe if you hire a worker as a subcontractor through a third party.
As far as the IRS is concerned, the company that benefits from a worker's services is responsible for the proper classification and treatment of that worker, regardless of whether the worker is engaged directly or as a subcontractor.

The Common Law Test

When a Legal Test Is Necessary 17

The Right of Control Is Key 19

Factors for Measuring Control 20

Making a Profit or Taking a Loss 21

Working on Site 22

Offering Services to the General Public 22

Right to Fire 22

Furnishing Tools and Materials 22

Method of Payment 23

Working for More Than One Business 23

Continuing Relationship 24

Investment in Equipment or Facilities 24

Business or Traveling Expenses 25

Right to Quit 25

Instructions 25

Sequence of Work 27

Training 27

Services Performed Personally 28

Hiring Assistants 28

Set Working Hours 29

Working Full Time 29

Oral or Written Reports 29

Integration Into Business 29

Skill Required 30

Worker Benefits 31

Tax Treatment of the Worker 32

Intent of the Hiring Firm and Worker 32

Custom in the Trade or Industry 32

Evaluating a Worker's Status.. 33

    Asking the Right Questions 33

    Interpreting Your Answers...36

A worker does not become an independent contractor simply because you say so. Courts and government agencies will determine the worker's status by applying a legal worker classification test. There are all sorts of classification tests, and we will introduce you to them in this book. In this chapter, however, we take a close look at the most frequently used test: the common law test. Many agencies use this test or some form of it.

One of the most well-known agencies that uses the common law test is the IRS. Because the IRS test is so important, you'll find an entire chapter—Chapter 3—on that particular use of the common law test. If you are looking for guidance on applying the IRS test or classifying a worker for federal payroll tax purposes, skip ahead to Chapter 3.

When a Legal Test Is Necessary

As discussed at the beginning of this book, ICs are people who are in business for themselves. Sometimes it's very easy to tell if workers are in business for themselves and, therefore, should be classified as independent contractors.

> **EXAMPLE 1:** You start a restaurant and contract with IBM to install a computer system for your business. There is no way IBM will be viewed as your employee. You need not worry about paying employment taxes for IBM's workers. That's IBM's problem. IBM is clearly an established independent business. Not even the most hard-nosed IRS auditor would question this.

> **EXAMPLE 2:** You hire several people to wait tables in your restaurant and pay them salaries, benefits, and so forth. It is clear that a typical waitperson in a restaurant is not running an independent business. He or she is an employee of the restaurant. The restaurant owner— that is, you—is the one in business.

In cases like these, it's so clear that the worker is—or is not—an independent businessperson that there really is no need to apply any specific legal test to determine the worker's status. In many other cases, however, the issue is not quite so clear. Things can be especially foggy when workers perform specialized services by themselves—that is, without the help of assistants.

> EXAMPLE: Instead of hiring IBM, you hire a computer consultant named Mike to install your computers. Mike has no employees and performs all the work for you personally. He spends several months working on your computers. It's difficult to say for sure that Mike was in business for himself while he worked for you.

Every worker in America falls somewhere on a continuum. At one end are workers who are clearly employees. At the other are those who are clearly ICs. But in between these two extremes, there is a vast ground where work relationships have some elements of employment and some elements of independence. It is in this uncertain terrain that problems with the IRS, state tax authorities, unemployment compensation authorities, and other agencies can occur.

Courts and agencies have developed detailed legal tests to determine the status of workers in this middle ground. The purpose of these tests is to give both businesses and government agencies some objective and understandable basis for classifying workers. Unfortunately, however, these tests often don't provide a clear answer about how to classify a worker.

The common law test—also called the right to control test—is the legal test most frequently used to determine worker status. This is the test used by:

- the IRS (see Chapter 3 for a detailed examination of how the IRS applies the common law test)
- unemployment compensation insurance agencies in many states (see Chapter 5 to find out which states use the common law test)
- workers' compensation insurance agencies in many states (see Chapter 6 to find out which states' workers' compensation agencies use the common law test)
- courts, to resolve copyright ownership disputes (see Chapter 9 for more information about intellectual property issues), and
- many federal regulatory agencies, such as the National Labor Relations Board (which enforces federal laws regarding unions).

> ## Other Tests for IC Status
>
> The common law test isn't the only test used to determine worker status. Two other tests are used by some government agencies:
> - **The Economic Reality Test.** Under this test, workers are employees if they are economically dependent upon the businesses for which they render services. (See Chapter 8 for a detailed discussion of this test.)
> - **ABC Test.** About half the states use a special statutory test, called the ABC test, to determine whether workers are ICs or employees for purposes of unemployment compensation. This test focuses on just a few factors. (This test is covered in detail in Chapter 5.)

The Right of Control Is Key

The common law test is based on a very simple notion: Employers have the right to tell their employees what to do. The employer may not always exercise this right—for example, if an employee is experienced and well trained, the employer may not feel the need to closely supervise him or her—but the employer has still the right to do so.

Under the common law test, workers are employees if the people for whom they work have the right to direct and control the way they do their jobs—both the final results and the details of when, where, and how the work is performed.

> **EXAMPLE:** Mary takes a job as a hamburger cook at the local Acme-Burger. AcmeBurger personnel carefully train her in how to make an AcmeBurger hamburger, including the type and amount of ingredients to use, the temperature at which the hamburger should be cooked, and so forth. Once Mary starts work, AcmeBurger managers closely supervise how she does her job.
>
> Virtually every aspect of Mary's behavior on the job is under AcmeBurger control—including what time she arrives at and leaves work, when she takes her lunch break, what she wears, and the

sequence of the tasks she must perform. If Mary proves to be an able and conscientious worker, her supervisors may not look over her shoulder very often, but they have the right to do so at any time. Mary is AcmeBurger's employee.

In contrast, when you hire an IC, you hire an independent businessperson. A business typically does not have the right to control the way an independent businessperson—an IC—performs agreed-upon services. Its control is limited to accepting or rejecting the final results.

EXAMPLE: AcmeBurger develops a serious plumbing problem. AcmeBurger doesn't have any plumbers on its staff, so it hires Plumbing by Jake, an independent plumbing repair business owned by Jake. Jake looks at the problem and gives an estimate of how much it will cost to fix. The manager agrees and Jake and his assistant commence work. The manager doesn't give Jake any instructions on how to fix the plumbing problem—the manager just wants it resolved.

In a relationship of this kind, where Jake is clearly running his own business, it's virtually certain that AcmeBurger does not have the right to control the way Jake performs his services. Its control is limited to accepting or rejecting the final result. If AcmeBurger doesn't like the work Jake has done, it can refuse to pay him.

Factors for Measuring Control

The difficulty in applying the common law test lies in figuring out whether a business has the right to control its workers. Government auditors can't look into your mind to see whether the right to control exists. They must rely primarily on indirect or circumstantial signs of control or lack of it—for example, whether you provide a worker with tools and equipment, pay by the hour, or have the right to fire him or her. This is what you'll have to answer questions about if you're audited.

To evaluate whether a worker passes muster as an IC, you need to examine these factors. The fact that you may know in your heart that you do not control a worker is not sufficient. What matters is how your relationship with the worker appears to a government auditor who doesn't know either of you.

Government auditors examine a number of different factors to determine whether a hiring firm has the right to control a worker. The following list includes virtually every factor any auditor might consider. No agency uses all 25 of these factors; instead, an agency may use anywhere from four to 20 from this list. Which factors are used by which agencies is discussed in later chapters.

You don't need to memorize this list. It's included so that you can refer to it if you need it. There's no magic number of factors that make a worker an IC or an employee. You need to look at the big picture. You may look at the list and find that so many factors weigh in favor of either IC status or employee status that you can feel secure in your classification. In other cases, you may go through the list and still feel like you don't know how to classify the worker. When that happens, consider consulting an expert, such as an accountant or attorney, for assistance.

Making a Profit or Taking a Loss

Employees are typically paid for their time and labor and don't have to pay business expenses. They earn the same salary regardless of how the work is performed.

In contrast, ICs can earn a profit or suffer a loss from their work. They make money if their businesses succeed, but risk going broke if they fail. Whether ICs make money depends on how well they use their ingenuity, initiative, and judgment in conducting their business.

A worker who has an opportunity to make a profit or suffer a loss based on the work being performed looks more like an independent contractor. This is an extremely important factor, particularly for highly skilled workers like doctors and lawyers, who ordinarily aren't supervised closely or given detailed instructions on how to do their work.

Working on Site

Employees must work where their employers tell them, usually on the employer's premises. ICs are often able to choose where to perform their services. Thus, the fact that a worker does the job at your workplace may weigh in favor of an employment relationship, especially if the work could be done elsewhere. A person who works at your place of business is physically within your direction and supervision. If the person can choose to work off site, you obviously have less control. However, today many employees work at home, at least part of the time. Thus, this factor is never determinative by itself.

Offering Services to the General Public

Employees usually offer their services solely to their employers; ICs ordinarily make their services available to the public. Thus, if the worker advertises or works for people other than you, this tends to show that he or she is an IC.

Right to Fire

Unless an employee has an employment contract, he or she typically can be fired by the employer at any time, for any reason that is not illegal. An IC's relationship with a hiring firm can be terminated only according to the terms of his or her agreement. If you have a right to fire a worker at any time for any reason or for no reason at all, government auditors may conclude that you have the right to control that worker. The ever-present threat of dismissal could pressure a worker to follow your instructions and otherwise do your bidding. Thus, the right to fire weighs in favor of employee status.

Furnishing Tools and Materials

Employers ordinarily give employees all the tools and materials necessary to do their jobs. ICs typically furnish their own tools and materials.

The fact that a hiring firm furnishes tools and materials, such as computers and construction equipment, tends to show control because the firm can determine which tools the worker is to use and, at least to some extent, how to use them. In most circumstances, then, the furnishing of tools and materials by the hiring firm weighs in favor of employee status.

> **TIP**
> **Sometimes, tools don't matter.** ICs may have to use a hiring firm's tools or materials. For example, a computer consultant may have to perform work on your company's computers. In such a situation, the fact that you provided the tools should be irrelevant.

Method of Payment

Employees are usually paid by unit of time—for example, by the hour, week, or month. In such a situation, the employer assumes the risk that the services provided will be worth what the worker is paid. To protect its investment, the employer demands the right to direct and control the worker's performance. In this way, the employer makes sure it gets a day's work for a day's pay.

ICs typically earn a flat rate for a project. The IC will have to make sure that the agreed-upon amount will adequately compensate for the time and money spent on the project. The IC, then, will control how the work gets done. Thus, payment by the job or on a straight commission generally weighs in favor of IC status.

In many professions and trades, however, ICs are customarily paid by unit of time. For example, lawyers, accountants, and psychiatrists typically charge by the hour. Where this is the general practice, the method-of-payment factor won't be given great weight.

Working for More Than One Business

Many employees have more than one job at a time. However, employees owe a duty of loyalty to their employers—that is, employees cannot engage in activities that harm or disrupt the employer's business. This restricts employees' outside activities. For example, an employee ordinarily wouldn't be permitted to take a second job with a competitor of the first employer. An employee who did so would be subject to dismissal.

ICs are generally subject to no such restrictions. They can work for as many clients or customers as they want. Having more than one client or customer at a time is very strong evidence of IC status. People who work for several firms at the same time are generally ICs because they're not under the control of any one of them.

Continuing Relationship

Although employees can be hired for short-term projects, this type of relationship is more typical of ICs. An employee usually works for the same employer month after month, year after year, sometimes decade after decade. Such a continuing relationship is one of the hallmarks of employment. Indeed, one of the main reasons businesses hire employees is to have workers available on a long-term basis.

ICs, on the other hand, come and go. A business hires an IC for a project, and the relationship ends when the work is done.

Investment in Equipment or Facilities

A worker who makes a significant investment in equipment and facilities to perform services is more likely to be considered an IC. By making such a financial investment, the worker risks taking a loss if the business isn't profitable. Also, the worker isn't dependent on someone else to provide the tools and facilities needed to do the work. Owning the tools and facilities also signals that the worker has the right to decide how and when to use them.

On the other hand, a worker who depends on the hiring business to provide equipment and facilities looks more like an employee than an IC.

This factor includes equipment and premises necessary for the work, such as office space, furniture, and machinery. It does not include tools, instruments, and clothing commonly provided by employees in their trade— for example, uniforms or hand tools that are commonly provided by the employees themselves. Nor does it include education, experience, or training.

Some types of workers typically provide their own inexpensive tools. For example, carpenters may use their own hammers and accountants their own calculators. Providing such inexpensive tools doesn't show that a worker is an IC. But a worker who provides his or her own $3,000 computer or $10,000 lathe is more likely to be classified as an IC.

Business or Traveling Expenses

The fact that a business pays a worker's expenses points to employee status. To be able to control such expenses, the employer must retain the right to regulate and direct the worker's actions.

On the other hand, a person who is paid per project and has to pay expenses out of pocket is more likely to be viewed as an IC. Any worker who is accountable only to himself or herself for expenses has more freedom to decide exactly how to do the job.

Of course, some ICs typically bill their clients for certain expenses. For example, accountants normally bill clients for travel, photocopying, and other incidentals. This fact alone doesn't make them employees, because their clients don't control the way they work.

Right to Quit

Employees normally work "at will." This means they can quit whenever they want to without incurring liability, even if it costs the employer substantial money and inconvenience.

ICs usually agree to complete specific jobs. If they don't complete the jobs, they are legally responsible for making good on any losses they cause to the hiring businesses.

Instructions

Employers have the right to give their employees oral or written instructions about when, where, and how they work. Businesses generally don't give independent contractors these sorts of instructions.

This can be a difficult factor to evaluate because it focuses on a business's right to give instructions, not on whether instructions were actually given.

If a worker is running an independent business and you are just one client or customer among many, you probably don't have the right to give him or her instructions about how to perform the services. Your right is usually limited to accepting or rejecting the final results.

> **EXAMPLE:** Art goes to Joe's Tailor Shop and hires Joe to make him a suit. Art chooses the fabric and style of the suit, but it's up to Joe to decide how to make the suit. When the suit is finished, Art can refuse to pay for it if he thinks it isn't made well. Joe is an independent contractor. If Art had presumed to tell Joe how to go about cutting the fabric and stitching the suit together, Joe would probably have kicked him out of his shop and gone on to his next customer.

On the other hand, you probably have the right to give instructions to workers who aren't running an independent business and who are largely or solely dependent upon you for their livelihood.

> **EXAMPLE:** Joe the tailor abandons his own tailor shop when he's hired to perform full-time tailoring services for Acme Suits, a large haberdashery chain. Joe is completely dependent upon Acme for his livelihood. Acme managers undoubtedly have the right to give Joe instructions, even if they don't feel the need to do so because he is such a good tailor.

A hiring firm may give an IC detailed guidelines as to the end results to be achieved. For example, a software programmer may be given highly detailed specifications describing the software programs to develop, or a building contractor may be given detailed blueprints showing precisely what the finished building should look like. Because these instructions relate only to the end results to be achieved, not how to achieve them, they don't make the workers employees.

> ## Highly Skilled Professionals Need Few Instructions
>
> Highly trained professionals, such as doctors, accountants, lawyers, engineers, and computer specialists, may require very little, if any instruction on how to perform their services. In fact, it may be impossible for the hiring firm to instruct the worker on how to perform the services in question because it lacks the knowledge and skill to do so. Nevertheless, highly skilled professionals can be employees. In cases involving highly trained pros, the IRS and courts place greater emphasis on the business relationship between the parties.

Sequence of Work

Employees may be required to perform services in the order or sequence set for them by the employer. ICs decide for themselves the order or sequence in which they work.

The sequence-of-work factor is closely related to the right to give instructions. If a person must perform services in the order or sequence set by the hiring firm, that requirement shows that the worker isn't free to use discretion, but must follow established routines and schedules.

Often, because of the nature of the occupation, the hiring business doesn't bother to require that tasks be done in a particular order. It is sufficient to show control, however, if the hiring firm retains the right to do so. For example, a salesperson who works on commission usually has some latitude in mapping out work activities. But one who hires such a salesperson normally has the discretion to require him or her to report to the office at specified times, follow up on leads, makes sales calls on a particular route or schedule, and so on. Such requirements interfere with and take precedence over the salesperson's own routines or plans. They indicate control by the hiring business and employee status for the salesperson.

Training

Employees may receive training from their employers. ICs ordinarily receive no training from those who purchase their services.

Training may be accomplished by teaming a new worker with a more experienced one, by requiring attendance at meetings or seminars, or even by correspondence. Training shows control because it indicates that the employer wants the services performed a particular way. This is especially true if the training occurs periodically or at frequent intervals.

ICs are usually hired precisely because they don't need any training. They possess special skills that the hiring firm's employees do not.

Services Performed Personally

Employees are required to perform their services on their own—that is, they can't get others to do their jobs for them. ICs ordinarily are not required to render services personally; for example, they can hire their own employees or even other ICs to do the work.

Ordinarily, when you hire an IC, he or she has the right to delegate all or part of the work to others without your permission. This is part and parcel of running a business. For example, if you hire an accountant to prepare your tax return, the accountant normally has the right to have assistants do all or part of the work under his or her supervision.

Requiring someone you hire to perform the services personally indicates that you want to control how the work is done, not just the end results. If you were just interested in end results, you wouldn't care who did the work—you'd just make sure the work was done right when it was finished.

Hiring Assistants

Employees hire, supervise, and pay assistants only at the direction of their employers. ICs, on the other hand, hire, supervise, and pay their own assistants without input from the hiring businesses.

Government auditors are usually impressed by the fact that a worker hires and pays his or her own assistants. This is something employees simply do not do and is strong evidence of IC status because it shows risk of loss if the worker's income doesn't meet payroll expenses.

Set Working Hours

Employees ordinarily have set hours of work. ICs are masters of their own time. They ordinarily set their own work hours.

Working Full Time

An employer might require an employee to work full time. ICs are free to work when and for whom they choose, and usually have the right to work for more than one client or customer at a time.

Oral or Written Reports

Employees may be required to submit regular oral or written reports to their employers regarding the progress of their work. ICs are generally not required to submit regular reports; they are responsible only for end results.

Submitting reports shows that a worker is compelled to account for individual actions. Reports are an important control device for an employer. They help the employer determine whether directions are being followed and whether new instructions should be issued.

Regular reports that enable an employer to keep track of workers' day-to-day performance indicate employee status. But it's quite common for ICs to make infrequent interim reports to hiring firms when they are working on long or complex projects. Such reports are typically tied to specific completion dates, timelines, or milestones written into the contract. For example, a building contractor may be contractually required to submit a report when each phase of a complex building project is completed.

Integration Into Business

Employees typically provide services that are an integral part of the employer's day-to-day operations. In contrast, an independent contractor's services are usually outside of what the business does to earn money.

Integration in this context has nothing to do with race relations. It simply has to do with whether the worker is a regular part of the hiring firm's overall operations. According to most government auditors, the hiring firm would likely exercise control over integrated workers because they are so important to the success of the business.

> EXAMPLE: Fry King is a fast food outlet. It employs 15 workers per shift who prepare and sell the food. Jean is one of the workers on the night shift. Her job is to prepare all the French fries for the shift. Fry King would likely go out of business if it didn't have someone to prepare the French fries. French fry preparation is a regular or integral part of Fry King's daily business operations.

On the other hand, ICs generally have special skills that the hiring firm calls upon only sporadically.

> EXAMPLE: Over the course of a year, Fry King hires a painter to paint its business premises, a lawyer to handle a lawsuit by a customer who suffered from food poisoning, and an accountant to prepare a tax return. All of these tasks may be important or even essential to Fry King, but they aren't a part of its regular, daily operations of selling fast food.

Skill Required

Workers whose jobs require a low level of skill and experience are more likely to be employees. Workers with jobs requiring specialized skills are more likely to be ICs.

The skill required to do a job is a good indicator of whether the hiring firm has the right to control the worker. This is because you are far more likely to have control over the way low-skill workers do their jobs than over the way high-skill workers do theirs.

For example, if you hire an experienced repair person to maintain an expensive and complex photocopier, it's doubtful that you know enough about photocopiers to supervise the work or tell the repair person what to do.

This isn't the case, however, when you hire a person to do a job that doesn't require highly specialized skills or training, such as answering telephones. You are likely to spell out the details of how the work should be done and are certainly capable of supervising the worker. Workers in such occupations generally expect to be controlled by the person who pays them—that is, they expect to be given specific instructions as to how to work, be required to work during set hours, be provided with tools and equipment, and so forth.

For these reasons, highly skilled workers are far more likely to be ICs than low-skill workers. However, not all high-skill workers are ICs. Corporate officers, doctors, and lawyers, for example, can be employees just like janitors and other manual laborers if they are subject to the control of the business that hires them.

> **EXAMPLE:** Dr. Smith leaves his lucrative solo medical practice to take a salaried position teaching medicine at the local medical school. When Smith ran his own practice, he was an IC in business for himself. He paid all the expenses for his medical practice and collected all the fees. If the expenses exceeded the fees, he lost money. As soon as Smith took the teaching job, he became an employee of the medical school. The school pays him a regular salary and provides him with employee benefits, so he has no risk of loss as he did when he was in private practice. The school also has the right to exercise control over Smith's work activities—for example, requiring him to teach certain classes. It also supplies an office and all the equipment Smith needs. He is no longer in business for himself.

Worker Benefits

Employees usually receive benefits such as health insurance, sick leave, pension benefits, and paid vacation. ICs ordinarily receive no such workplace benefits.

If you provide workers with employee benefits, it's only logical for courts and government agencies to assume that you consider them to be your employees, subject to your control. You'll have a very hard time convincing anyone that a person you provide with employee benefits is not your employee.

Tax Treatment of the Worker

Employees have federal and state payroll taxes withheld by their employers and remitted to the government. ICs pay their own taxes.

Treating a worker as an employee for tax purposes—that is, remitting federal and state payroll taxes for the worker—is very strong evidence that you believe the worker to be your employee and that you have the right to exercise control over him or her. Indeed, one court has ruled that paying federal and state payroll taxes for a worker is a virtual admission that the worker is an employee under the common law test. (*Aymes v. Bonelli*, 980 F.2d 857 (2d Cir. 1992).)

Intent of the Hiring Firm and Worker

If it appears that the business and the worker honestly intended to create an IC relationship, it's likely that the business would not believe it had control, nor attempt to exercise control, over the worker. One way to establish intent to create an IC relationship is for both parties to sign an independent contractor agreement. (See Chapter 12 for guidance on creating such an agreement.)

On the other hand, if it appears that you never intended to create a true IC relationship and merely classified the worker as an IC to avoid an employer's legal obligations, the worker will likely be considered an employee.

Custom in the Trade or Industry

The custom of classification in the trade or industry involved is an important part of the analysis. If a particular type of work is usually performed by employees, workers in that field are more likely to be classified as such.

> EXAMPLE: The longstanding custom among logging companies in the Pacific Northwest is to treat tree fellers—people who cut down trees—as ICs. They are customarily paid by the tree, receive no employee benefits, and are free to work for many logging companies,

not just one. None of the logging companies withhold or pay federal or state payroll taxes for tree fellers. The fellers pay their own self-employment taxes. This longstanding custom (among other factors) is strong evidence that the workers are ICs.

Evaluating a Worker's Status

As you've seen, there are lots of factors to consider in determining whether an employer has the "right of control" over a worker. Having such factors laid out in one place can be a big help.

Asking the Right Questions

The following questionnaire is based on the *Employment Determination Guide*, a tool produced by the California Employment Development Department. It provides an outstanding guide to applying the common law right-of-control test for determining worker status.

Questions 1 through 3 are significant. A "yes" answer to any of them is a strong indication that the worker is an employee, and that you are at risk if you classify him or her as an independent contractor.

"No" answers to each of Questions 4, 5, and 6 indicate that the individual is not in business for him- or herself and would therefore normally be an employee.

The greater the number of "yes" answers to Questions 7 through 13, the greater the likelihood the worker is performing services as an employee. However, no factor by itself is deciding. All must be considered and weighed to determine which type of relationship exists.

Employment Status Questionnaire			
Question	Yes	No	Explanation
1. Do you instruct or supervise the person while he or she is working?			Independent contractors are free to do jobs in their own way, using methods they choose. A person or firm engages an independent contractor for the job's end result. A worker who is required to follow company procedure manuals or is given specific instructions on how to perform the work is normally an employee.
2. Can the worker quit or be discharged (fired) at any time?			The right to fire workers without notice indicates that you have the right to control them. Independent contractors are engaged to do specific jobs and cannot be fired before the job is complete unless they violate the terms of the contract. Likewise, they aren't free to quit and walk away until the job is complete. For example, if a shoe store owner hires an attorney to review his lease, the attorney gets paid only after satisfactory completion of the job.
3. Is the work being performed part of your regular business?			Work that is a necessary part of the regular trade or business is normally done by employees. For example, a sales clerk sells shoes in a shoe store—the shoe store owner wouldn't be able to operate without such clerks. On the other hand, a plumber engaged to fix the pipes in the bathroom of the store performs a service on a one-time or occasional basis, and the work isn't an essential part of the business's purpose. A certified public accountant engaged to prepare tax returns and financial statements for the business is another example of an independent contractor.
4. Does the worker have a separately established business?			The fact that workers hold themselves out to the public as available to perform services similar to those they perform for you suggests that they are independent contractors. Independent contractors are free to hire employees and assign the work to others in any way they choose. They have the authority to fire their employees without your knowledge or consent. They normally advertise their services in newspapers and other publications, and seek new customers through the use of business cards and the like.

Employment Status Questionnaire (continued)			
Question	Yes	No	Explanation
5. Is the worker free to make business decisions that affect his or her ability to profit from the work?			People are normally independent contractors when they are free to make business decisions that impact their bottom line. Their decisions involve real economic risk, not just the risk of not getting paid. These decisions normally involve the acquisition, use, or disposition of equipment, facilities, and stock-in-trade that are under their control. Other examples of decisions that directly affect profit include selection of the amount and type of advertising for the business, the priority of assignments, and the types and amounts of insurance coverage for the business.
6. Does the individual have a substantial investment that would subject him or her to a risk of financial loss?			Independent contractors normally have an investment in the items needed to complete their tasks. To the extent necessary for the specific type of business, independent contractors provide their own business facility, such as office space. If they don't earn enough to recoup their investment, they suffer a loss.
7. Do you have employees who do the same type of work?			The fact that the work is basically the same as what your employees do indicates that the worker is an employee. This is true even if the work is being done on a one-time basis. For instance, consider the hiring of a worker on a temporary basis to handle an extra workload or replace an employee who is on vacation. This worker is a temporary employee, not an independent contractor. (Note: If you contract with a temporary agency to provide a worker, the worker is normally an employee, but may be the employee of the temporary agency. See Chapter 10.)
8. Do you furnish the tools, equipment, or supplies used to perform the work?			Independent businesspeople furnish the tools, equipment, and supplies needed to perform the work.
9. Is the work considered unskilled or semi-skilled labor?			Workers who are considered unskilled or semiskilled are the type the law is meant to protect and are generally employees.

Employment Status Questionnaire (continued)			
Question	Yes	No	Explanation
10. Do you provide training for the worker?			In skilled or semiskilled work, independent contractors usually don't need training. If training is required to do the task, it is an indication that the worker is an employee.
11. Is the worker paid a fixed salary, an hourly wage, or based on a piece-rate basis?			Independent contractors agree to do a job and bill for the service performed. Payments to them for labor or services are made upon the completion of the project or completion of the performance of specific portions of it.
12. Did the worker previously perform the same or similar services for you as an employee?			The worker having previously performed the same or similar services for you as an employee is an indication that he or she is still an employee.
13. Does the worker believe that he or she is an employee?			Although belief of the parties is not controlling, it is a factor to consider when making an employment- or-independent-contractor determination. When both the worker and principal believe the former is an independent contractor, there is an argument to support an independent contractor relationship between the parties.

Interpreting Your Answers

The results of this questionnaire depend in large part on the services and the business, and will vary from one situation to another. There may be some factors leaning toward employment and some toward independence. Questions 1 through 6 relate directly to the presence or absence of direction and control. The answers to Questions 7 through 13, when joined with other evidence, may carry greater weight when indicating the presence or absence of direction and control.

Consider the four following response patterns:

- If the answers to Questions 1 through 3 are "no" and the answers to Questions 4 through 6 are "yes," there is an indication of independence. When this is the case, there are likely to be a number of "no" answers to Questions 7 through 13, supporting the independent contractor determination.

- If the answers to Questions 1 through 3 are "yes" and those to Questions 4 through 6 are "no," there is very strong indication that the worker in question is an employee. It's likely that there will be a number of "yes" answers to Questions 7 through 13, supporting the employee classification.

- If the answer to Question 1 or 2 is "yes" or the answer to any of Questions 4 through 6 is "no," employment is likely. At the very least, this pattern of answers makes the determination more difficult since the responses to Questions 7 through 13 will probably be mixed.

- If the answer to Question 3 is "yes" and the answer to question 4 is "no," there is a likelihood of employment. Given this pattern of answers, it's probable that the answers to Questions 5 and 6 will also be "no." When this happens, you may also see more "yes" answers to the last group of questions (7 through 13). This scenario would support an employment classification.

These four patterns don't illustrate all the combinations of answers that could result from use of this questionnaire. The more the answers vary from the above four patterns, the more difficult the classification will be. But if the answers resemble the first and second patterns, there is a lesser risk of misclassifying the worker.

How the IRS Classifies Workers

Four Steps to Classification Under the IRS Rules 40

Step 1: Check Statutory Independent Contractor Rules 46

    Direct Sellers.. 46

    Licensed Real Estate Agents 47

Step 2: Analyze the Worker Under the Common Law Test................................. 49

    Behavioral Control... 51

    Financial Control 53

    Relationship of the Parties... 56

    Applying the Test 59

    IRS and Court Rulings for Specific Occupations.. 61

Step 3: Check Statutory Employee Rules... 72

    Corporate Officers... 74

    Home Workers... 75

    Food, Beverage, and Laundry Distributors... 77

    Life Insurance Salespeople.. 78

    Business-to-Business Salespeople.. 79

Step 4: Check the Safe Harbor Rules... 80

    First Prong: Filing IRS Form 1099-MISC When Appropriate.................... 81

    Second Prong: Consistent Treatment.. 83

    Third Prong: Having a Reasonable Basis for IC Classification 88

    Limitations on Safe Harbor Protection... 96

P erhaps the most important reason to classify workers correctly is to make sure you abide by federal payroll tax rules. These rules generate the biggest taxes you will have to pay for a worker who is an employee—and impose the biggest fine if you misclassify a worker as an independent contractor.

The heavy hand in all of this is wielded by the U.S. Internal Revenue Service. This is the agency that enforces the federal payroll tax rules, and it is the agency that will audit you and levy fines against you if you violate those rules. If you learn only one set of rules for determining who is and who is not an independent contractor, the IRS test should be it.

This chapter explains how to determine whether a worker is an independent contractor or employee under the IRS rules. For state tests, see Chapters 5 and 6.

> **RESOURCE**
> **Get the IRS rules.** The IRS provides detailed information about federal payroll taxes and its employee/independent classification rules in Publication 15 and Publication 15-A, respectively.

Four Steps to Classification Under the IRS Rules

To determine whether a worker is an employee or independent contractor in the eyes of the IRS, you must go through the following four steps:

Step 1: Check to see whether the worker is a statutory independent contractor—that is, someone who is an IC simply because the law says so. Two categories of workers can be statutory ICs: direct sellers and licensed real estate agents. (See "Step 1: Check Statutory Independent Contractor Rules," below, for more information about statutory ICs.) If the worker is not a statutory independent contractor, go on to Step 2. If the worker is a statutory IC, you don't have to bother with the remaining steps. You can treat the worker as an IC, and you don't have to pay or withhold federal payroll taxes.

Federal Payroll Taxes: The Bane of Having Employees

Federal payroll taxes are one of the chief reasons that employees are more expensive than independent contractors. You must pay these taxes for each employee on your payroll. They include:

- **Social Security and Medicare taxes:** Also called employment or FICA taxes, they pay for the Social Security and Medicare systems. These consist of a 12.4% Social Security tax up to a ceiling amount that is adjusted each year for inflation. In 2014, the ceiling was $117,000. The 2.9% Medicare tax must be paid on all employee wages. Employers must pay half of these taxes themselves and deduct the other half from their employees' pay, then send all the money to the IRS. Thus, employers must pay a 6.2% Social Security tax up to the annual ceiling and a 1.45% Medicare tax on all their employees' wages.

- **Federal unemployment tax:** Also called FUTA, this is ordinarily an 0.8% tax on the first $7,000 employees are paid annually, or $56 per year per employee. Employers must pay FUTA taxes out of their own pockets. This tax must be paid if an employer (1) pays at least $1,500 in total wages to employees in any three-month period, or (2) has at least one employee during any day of a week during 20 weeks in a calendar year (the 20 weeks need not be consecutive).

- **Income tax withholding:** Employers don't have to contribute to their employees' income taxes, but they do have to withhold them from employees' paychecks and pay them to the IRS on their behalf—in effect, acting as unpaid tax collectors for the IRS.

You do not have to pay or withhold these taxes for independent contractors, who pay their own Social Security and Medicare taxes in the form of self-employment taxes and who pay income taxes directly to the IRS, usually in the form of quarterly estimated taxes.

IRS Circular E, *Employer's Tax Guide,* provides detailed information on federal payroll taxes. It is an outstanding resource that every employer should have. You can download a copy of the guide from the IRS website at www.irs.gov.

Step 2: Apply the IRS common law test to determine whether the worker is an IC or employee. (See "Step 2: Analyze the Worker Under the Common Law Test," below, for information about the IRS common law test.) If you are certain the worker should be classified as an IC under these rules, proceed to Step 3. If you think the worker should be classified as an employee under the common law test, go to Step 4.

Step 3: Check to see whether the worker is a statutory employee. (The section called "Step 3: Check Statutory Employee Rules," below, walks you through this process in more detail.) If the worker is not a statutory employee, and you determined that the worker could be classified as an independent contractor in Step 2, then you can stop. The worker is an IC, and you don't have to pay federal payroll taxes. If the worker is a statutory employee, go on to Step 4.

Step 4: If you think the worker should be classified as an employee under the common law test, or if the worker is a statutory employee, figure out whether the worker fits into the safe harbor rules. (See "Step 4: Check the Safe Harbor Rules," below, for guidance on applying those rules.) If so, you can treat the worker as an independent contractor. You don't have to pay or withhold federal payroll taxes. If the worker does not fit into the safe harbor rules, then the worker is an employee and you must pay federal payroll taxes.

Although the IRS is in charge of collecting federal payroll taxes, it's up to you to make the initial decision as to whether you must withhold and pay such taxes for workers. That is, you must decide whether each of your workers is an employee or an independent contractor and then pay your federal taxes accordingly.

Eventually, the IRS may review and audit your classifications. The IRS takes this enforcement task very seriously. If it determines that you wrongly classified workers as ICs, it will impose fines and penalties. (See Chapter 4 for more about IRS audits.)

Special Rules for Hiring Workers Who Are Not U.S. Citizens

Special federal tax withholding rules may apply to businesses that hire aliens (people who are not U.S. citizens) to perform services. The rules depend on whether the worker is a resident or nonresident alien. A resident alien is a person who has a "green card" (an immigrant visa) or resided in the United States for more than 183 days of the current year and each of the previous two years. Nonresident aliens are those who don't meet these requirements.

Whether resident and nonresident aliens are employees or ICs for federal tax purposes is determined under the same rules that apply to any other worker. If a resident alien is an employee, he or she must be treated like any other employee. Likewise, you may treat a resident alien who qualifies as an IC just as you would any other IC.

However, the rules are different for nonresident aliens. Even if a nonresident alien qualifies as an IC, you'll have to withhold 30% of his or her pay and remit it to the IRS unless the worker qualifies for an exemption from this withholding rule. Fortunately, many nonresident alien ICs qualify for an exemption because they are from countries with which the United States has signed a tax treaty eliminating the withholding requirement if certain conditions are met. A list of these treaty countries and their requirements can be found in IRS Publication 515, *Withholding of Tax on Nonresident Aliens and Foreign Entities.*

To obtain the exemption, the nonresident IC must complete IRS Form 8233, *Exemption From Withholding on Compensation for Independent (and Certain Dependent) Personal Services of a Nonresident Alien Individual.* The hiring business must sign the form and send it to the IRS. For detailed instructions on withholding requirements for nonresident aliens, look at IRS Publication 515, *Withholding of Tax on Nonresident Aliens and Foreign Entities.*

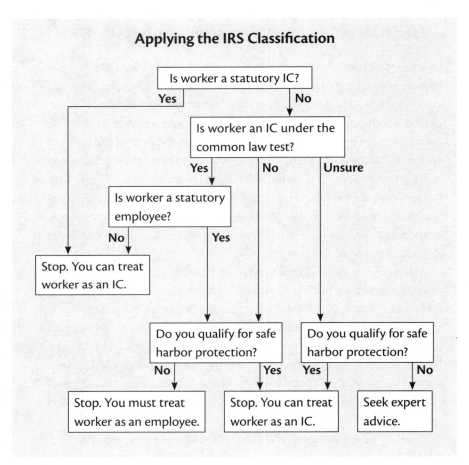

Asking the IRS for Determination of Worker Status: A Pointless Exercise?

A worker or business may ask the IRS for a determination as to whether a worker (or group of similar workers) is an employee or independent contractor by filing IRS Form SS-8, *Determination of Worker Status for Purposes of Federal Employment Taxes and Income Tax Withholding.* The four-page form requires the submitter to answer numerous questions and provide a detailed explanation of the work relationship between the business and the worker. The other party—whether worker or hiring firm—will be asked by the IRS to complete an SS-8 as well.

Asking the IRS for Determination of Worker Status: A Pointless Exercise? (continued)

An IRS technician will then make a written determination of whether the worker (or group of workers) is an employee or IC for IRS purposes. The determination is binding on the IRS, but applies only to the business and the worker it addresses.

If the IRS determines that the worker or workers involved were misclassified as ICs, the employer is instructed to file corrected employment tax returns and information returns showing that the worker and any other worker performing services under the same circumstances are treated as employees. The SS-8 procedure is not an audit, so no assessments or penalties are imposed.

This all sounds great in theory, but in practice, the SS-8 program has not worked out so well. It takes about one year for the SS-8 Unit to process the form—a time delay that will likely grow longer due to IRS budget problems. Moreover, over 70% of the time, the unit tells a business that files an SS-8 form to classify the worker as an employee. Workers were found to be ICs by the IRS only 3% of the time. This undoubtedly reflects the IRS's bias in favor of classifying as many workers as possible as employees.

A worker or business disagreeing with a determination may request reconsideration; the worker or business must provide additional information. However, the same group handles the reconsideration and almost all of the original determinations are upheld.

Under current procedures, no administrative appeal rights are allowed. Perhaps as a result, a 2013 study by the Treasury Inspector General for Tax Administration found that 19% of employers fail to comply with adverse determination rulings by the SS-8 Unit. The SS-8 Unit has little ability to enforce compliance after a determination ruling, but it can refer the case for an audit.

It's no wonder then that over 95% of all SS-8 forms are initially filed by workers, not hiring firms. Ultimately, filing a Form SS-8 is usually not worth the time and trouble for a hiring firm.

CAUTION

Get help if you need it. Because IRS penalties for misclassifying employees as ICs can be substantial, it's important to classify all workers correctly. If you're not sure how to classify a worker after reading this book, seek assistance from a tax expert.

Step 1: Check Statutory Independent Contractor Rules

To qualify as a statutory IC, the worker must be a direct seller or licensed real estate agent (as discussed below) and must meet two threshold requirements:

- The worker's pay must be based on sales commissions, not on the number of hours worked.
- There must be a written contract with the hiring business stating that the worker will not be treated as an employee for federal tax purposes. (See Chapter 12 for guidance on creating a written independent contractor agreement.)

The IRS calls these workers "statutory nonemployees," which is bureaucratese for ICs. The fortunate employers of these workers need not pay FICA or FUTA taxes or withhold federal income taxes.

> CAUTION
>
> **Don't forget state taxes.** These rules apply only to federal taxes—Social Security (FICA), federal unemployment tax (FUTA), and federal income tax withholding. You may still have to pay state payroll taxes and provide workers' compensation coverage depending on how the worker should be classified under the tests of the state involved. See Chapter 5 for information on state payroll taxes. See Chapter 6 for information on state workers' compensation issues.

Direct Sellers

Direct sellers (commonly referred to as door-to-door salespeople) sell consumer products to people in their homes or at a place other than an established retail store—for example, at swap meets. The products they sell include tangible personal property that is used for personal, family, or household purposes. These products include such things as vacuum cleaners, cosmetics, encyclopedias, gardening equipment, and similar consumer goods. They also include intangible consumer services or products such as home study courses and cable television services.

EXAMPLE: Larry is a Mavon Guy. He sells men's toiletries door to door. He is paid a 20% commission on all his sales. This is his only remuneration from Mavon. He has a written contract with Mavon stating that he will not be treated as an employee for federal tax purposes. Larry is a statutory independent contractor. Mavon need not pay FICA or FUTA, or withhold federal income taxes for Larry. It's up to Larry to pay his own self-employment taxes. Even if Larry would not qualify as an IC under the common law test, Mavon can safely treat him as an IC.

Direct sellers also include people who sell or distribute newspapers or shopping news. This is true whether they are paid by the publisher based on the number of papers delivered or they purchase newspapers from the publisher and then sell them and keep the money.

Licensed Real Estate Agents

Most states require real estate agents—called real estate salespeople in some states—to be licensed and work for licensed real estate brokers who are legally responsible for their actions. Real estate agents are usually paid on a straight commission basis. Whether they pay their own expenses and how much control the broker exercises over them varies from firm to firm.

Some real estate agents would probably qualify as employees under the common law test, while others would probably be ICs. Regardless of the common law test, however, they are statutory ICs if they meet the threshold requirements that we described above.

The broker need not pay FICA or FUTA, or withhold federal income taxes.

EXAMPLE: Mary is a licensed real estate agent who works for the Boldwell Canker real estate brokerage firm. Art, the real estate broker for whom she works, provides her with office space, pays most of her expenses, and exercises a good deal of control over her actions. However, Mary's pay is based solely on commission fees from properties she lists and sells. Boldwell Canker has a written agreement with Mary providing that she will not be treated as an employee for federal tax purposes.

Mary would probably qualify as Boldwell's employee under the common law test, but this doesn't matter because she is a statutory independent contractor. Boldwell need not pay FICA or FUTA, or withhold federal income taxes for Mary.

Fishing Boat Crews

Although they don't qualify as statutory nonemployees, members of a fishing boat crew are ICs for employment tax purposes (FICA and FUTA) if the boat's crew normally has fewer than ten members and the crew are paid:

- a share of the boat's catch of fish, or
- a share of the money from the sale of the catch if their pay depends on the amount of the catch—that is, they don't receive a guaranteed minimum amount. (26 U.S.C. § 31.3121(b)(20).)

Crew members will qualify as ICs even if they receive up to $100 per trip for taking on additional duties as a mate, engineer, or cook, for which additional cash payments are traditional in the fishing industry.

EXAMPLE: George, a commercial fishing boat owner, hires a crew of five for a fishing trip to the Grand Banks to catch cod. Each crew member gets a 10% share of the money earned from the sale of the catch. The crew members are ICs for employment tax purposes. This means that George need not pay or withhold Social Security, Medicare, unemployment, or federal income taxes for his crew.

For more information on tax issues affecting fishing boat crews, refer to the Fishing Tax Center at the IRS website at www.irs.gov/Businesses/Small-Businesses-&-Self-Employed/Fishing-Tax-Center.

Step 2: Analyze the Worker Under the Common Law Test

Unless a worker is a statutory IC (see above) or statutory employee (see below), the IRS uses a version of the common law test to determine a worker's status for federal payroll tax purposes. This test is also called the "right of control" test because it examines whether the hiring firm has the right to control the worker on the job. If so, the worker is an employee; if not, the worker is an IC.

Can a Worker Be an Employee and IC for the Same Firm?

The IRS says that, yes, a worker can be both an employee and independent contractor for the same hiring firm. The IRS Chief Counsel advised that a professional consultant working on two projects at same time for the same company could be classified as an employee on one project and IC on the other. In such cases, the IRS examines the work relationship separately for each service provided. If the worker qualifies as an IC for one project, the fact that he or she is an employee for the other project is not controlling. (IRS Info. Ltr. 01-0069.) Obviously, in such cases, the parties need to carefully document that the worker is being properly treated as an IC on the IC project.

The IRS looks at three areas to determine whether a hiring firm has the right to control a worker. These are:

- behavioral control on the job
- financial control, and
- the firm's relationship with the worker.

This section takes a closer look at these three areas. Be warned, however, that in close cases the analysis can be tricky. It is not necessary for all or even a majority of the factors to indicate IC status for a worker to be classified that way. All that is required is that the factors showing lack of control outweigh those showing control. There is no magic formula, and no one factor alone is enough to make a worker an employee or an IC. It can be difficult to decide which way the scales tip because not all the factors are

Primary IRS Common Law Factors		
	A worker will more likely be considered an IC if:	**A worker will more likely be considered an employee if:**
Behavioral control		
Factors that show whether a hiring firm has the right to control how a worker performs the specific tasks he or she has been hired to do	• You do not provide training. • You do not give the worker instructions on how to do the work. • You do not evaluate how the worker performs.	• You provide instructions that the worker must follow. • You give the worker detailed training. • You evaluate how the worker does the job.
Financial control		
Factors showing whether a hiring firm has a right to control a worker's financial life	• The worker has a significant investment in equipment and facilities. • The worker pays his or her own business or travel expenses. • He or she makes his or her services available to the public. • He or she is paid by the job. • He or she has opportunity for profit or loss.	• You provide equipment and facilities free of charge. • You reimburse the worker's business or traveling expenses. • The worker makes no effort to market his or her services to the public. • You pay him or her by the hour or other unit of time. • He or she has no opportunity for profit or loss—for example, because you pay by the hour and reimburse all expenses.
Relationship of the worker and hiring firm		
Factors showing whether you and the worker believe he or she is an IC or employee	• You don't provide employee benefits such as health insurance. • You sign an IC agreement with the worker. • The worker performs services that are not a part of your regular business activities.	• You provide employee benefits. • You have no written IC agreement. • The worker performs services that are part of your core business.

equally important in every case. The instructions below explain how to apply this test.

Behavioral Control

Behavioral control refers to the extent of your right to tell the worker exactly how to do the job. Giving instructions, providing training, and using evaluation systems are all strong evidence that you have the right to control a worker on the job, which makes the worker look like an employee in the IRS's eyes.

Giving Instructions

The single most important factor in the common law test is giving instructions. In the worker classification context, giving instructions means telling—or having the right to tell even if you don't exercise that right—a worker how to get the job done. Giving instructions includes telling a worker:

- when to do the work
- where to do the work
- what tools or equipment to use
- which other workers will help do the work
- where to purchase supplies or services
- what work must be performed by a specified person
- what routines or work patterns the worker must follow, and
- what order or sequence to follow in doing the work.

The more detailed the instructions, the more control the hiring firm exercises over the worker, and the more the worker looks like an employee. Requiring a worker to obtain prior approval before taking an action also constitutes giving instructions.

> EXAMPLE: Tanya is a truck driver who makes local deliveries for Zebra, Inc. She reports to the warehouse every morning. The warehouse manager tells Tanya which deliveries have to be made, how to load the cargo in the truck, what route to take, and the order in which items are to be delivered. These instructions strongly indicate that Tanya is Zebra's employee.

In contrast with instructions that tell a worker how to do the job, instructions about the end results a worker must achieve are perfectly consistent with IC status. Indeed, virtually every business instructs its ICs about what must be done—for example, when the work must be completed and what form the finished work product should take.

> **EXAMPLE:** John, a self-employed truck driver, receives a call from Acme, Inc., to make a delivery run from the Gulf Coast to the Texas Panhandle. John accepts the job and agrees to pick up the cargo the next morning. Upon arriving at the warehouse, John is given an address at which to deliver the cargo and is told that the delivery must be completed in two days. These instructions concern what must be done, not how it is to be done, and are therefore consistent with IC status.

The IRS distinguishes between giving instructions and giving suggestions. A suggestion about how work is to be performed does not constitute the right to control. For example, a dispatcher may suggest that an IC truck driver avoid a particular highway because of traffic congestion. If the worker is free to ignore the advice and use the highway anyway, then the dispatcher's comment is merely a suggestion and consistent with IC status. If complying with the statement is mandatory or if the worker would suffer adverse consequences for noncompliance, however, then the comment is, in fact, an instruction and consistent with employee status.

Similarly, businesses can require ICs to comply with government or industry body rules and regulations. However, if the firm establishes more stringent guidelines than those required by the government or industry body, the IRS will view them as instructions indicating employee status.

Highly skilled professionals, such as doctors, lawyers, accountants, engineers, and computer specialists, usually receive few instructions about how to perform their services. Firms that hire such workers frequently lack the knowledge necessary to give them such instructions. For this reason, in analyzing the status of professional workers, the IRS focuses more on the other two areas described in this section: financial control and evidence concerning the relationship of the parties.

> ⚠️ **CAUTION**
> **Wearing a uniform doesn't make a worker an employee.** In the past, the IRS viewed a firm's edict that workers wear company uniforms or put company logos on their vehicles as a sign of employee status. Not any more. The IRS now recognizes that concerns about safety cause many businesses to tell their customers that the workers will be in a company uniform or have a company logo on their vehicles. Requiring a worker to don a uniform or logo for security purposes is no longer viewed as an indicator of employee status.

Training

Providing a worker with periodic or ongoing training about procedures to be followed or methods to be used on a job is a strong indicator of employee status. However, the IRS now recognizes that not all training rises to this level. You may provide an IC with a short orientation or information session about your company policies, new product line, or new government regulations without jeopardizing the worker's IC status. Moreover, workers who attend voluntary training programs for which they aren't paid don't automatically lose their status as ICs.

Evaluations

Like instructions, evaluation systems are used by virtually all businesses to monitor the quality of work performed by workers. It is permissible to evaluate the quality of an IC's final work product. However, you should not use an evaluation system to measure compliance with performance standards concerning the details of how the work is performed. This shows you're trying to control how the worker does his or her job, which indicates employee status.

Financial Control

The second area of the IRS common law test looks for evidence of financial control—that is, whether you have the right to control how the worker conducts business. The factors that the IRS usually looks at are whether the worker:

- has a significant investment in equipment or facilities
- is not reimbursed for expenses
- makes his or her services available to the public
- is paid by the job or by some other method, and
- has an opportunity for profit or loss.

Significant Investment

Having a significant investment in equipment or facilities is a good indication that a worker is an IC. The IRS has not set any precise dollar amounts on what constitutes a significant investment, but says that the investment must have substance.

An IC doesn't necessarily have to purchase costly equipment to have a significant investment—he or she may rent or lease it instead of buying it. Indeed, an IC can even lease equipment from the hiring firm. As long as the worker pays fair rental value, the lease may constitute a significant investment.

> EXAMPLE: Charlene is a backhoe operator for the Yippee Distributing Company. She rents a $75,000 backhoe from Yippee for $1,000 per month—its fair rental value—and pays for liability insurance and regular maintenance on the backhoe. This is clearly a significant investment and helps establish that Charlene is an IC.

The IRS recognizes that many ICs do work that does not require large expenditures for equipment. For example, an IC consultant or salesperson may only require a telephone and home office. For this reason, a significant investment in equipment or facilities is not required for IC status.

Business Expenses

The more expenses a worker has that a hiring firm does not reimburse, the more the opportunity for profit or loss, and the more control the worker has over his or her financial life, the more likely the worker is to be an IC. Typical expenses for people who work for themselves include:

- rent and utilities
- advertising
- training
- wages of assistants
- licensing and professional dues
- insurance
- postage and delivery
- repairs and maintenance
- supplies
- travel, and
- inventory.

Of course, many ICs have many or all of their expenses reimbursed by their clients. The IRS recognizes this and does not view it as strong evidence of employee status.

IRS auditors are particularly impressed by fixed ongoing costs that a worker incurs regardless of whether work is currently being performed—for example, office rent and salaries for assistants.

Making Services Available to the Public

Unlike employees, ICs' economic prosperity usually depends on their ability to get new business. To do so, ICs often advertise, maintain a visible business location, and generally make themselves available for work in the relevant market. Evidence of such behavior—for example, a copy of a yellow pages ad—indicates IC status.

The IRS recognizes that not all ICs need to advertise or make themselves available to new clients, however. For example, many ICs obtain work by word of mouth without the need for advertising. Moreover, an IC who has negotiated a long-term contract may find advertising unnecessary and may be unavailable to work for others for the duration of a contract. For these reasons, the fact that a worker does not advertise, have a visible business location, or make him or herself available to new clients does not necessarily indicate employee status.

Method of Payment

Paying a worker a flat fee for an entire job is very strong evidence that he or she is an IC, especially if the worker's expenses aren't reimbursed. The worker runs the risk of losing money if the job takes longer or expenses are higher than anticipated. But he or she can also earn a windfall if the work can be done quickly with lower expenses than anticipated. Such an opportunity to earn a profit or suffer a loss is a hallmark of IC status.

Unfortunately, many ICs refuse to be paid by the job. Many insist on hourly, daily, or weekly payment. A worker who is compensated in this manner is guaranteed a return for his or her labor and has little risk of loss. This is generally evidence of employee status.

The IRS does recognize that some ICs—lawyers, for example—are typically paid by the hour instead of a flat fee. It's better not to pay an IC an hourly wage, but you can probably get away with it if it's a common practice in the IC's line of business.

Paying a worker on a commission basis is a neutral factor—that is, it doesn't cut one way or the other when the IRS evaluates a worker's classification.

Realization of Profit or Loss

The IRS believes that the ability to realize a profit or incur a loss is the strongest evidence that a worker is running an independent business. All the factors discussed above are relevant in determining whether a worker can earn a profit or suffer a loss. In addition, the IRS considers whether a worker is free to make business decisions that affect profit or loss.

Relationship of the Parties

The IRS lumps several additional factors under the title of relationship of the parties. What it's basically looking for is how you and the worker perceive your relationship to each other. The IRS figures that if you and a worker intended to create an IC relationship instead of an employee relationship, you probably won't have the right to control the worker. The IRS can't read your mind, so it looks for concrete evidence of your intent.

Factors That Are No Longer Important

Recognizing the changing nature of work in modern America, the IRS now de-emphasizes several factors that it used to consider crucial in determining worker status.

Working full time. It used to be considered the kiss of death for an IC to work full time for one client. IRS auditors saw this as very strong evidence of employee status. However, the IRS now recognizes that an independent contractor may work full time for one business because other contracts are lacking or because the contract requires exclusive effort.

Having multiple clients. The IRS used to view having multiple clients as strong evidence of IC status. Indeed, hiring firms would often encourage their ICs to work for others. It now says that having multiple clients is no longer useful evidence of IC status because many employees moonlight by working for a second employer.

Place of performance. Working on a hiring firm's premises used to be regarded as important evidence of employee status. No longer. The IRS now says that where the work is performed is largely irrelevant. However, this factor is still important in many state audits. For example, in half the states, you may be considered an employer for unemployment compensation purposes if a worker works at your place of business or another place you designate. (See Chapter 5 for information about state tests.)

Hours of work. The fact that a business requires a worker to perform services during certain hours used to be viewed as an indicator of employee status. Again, the IRS has changed its view. Some work must, by its nature, be performed at a specific time. On the other hand, the fact that a worker has flexible hours is not considered evidence of IC status because many employees now work flexible hours.

Right to fire or quit. The IRS no longer views an unfettered right to quit or fire as strong evidence of an employee relationship.

Use of Written IC Agreements

A written agreement describing the worker as an IC is good evidence that you and the worker intended to create an IC relationship. A written IC agreement, by itself, can never make a worker an IC, but it can be helpful in close cases. If the evidence is so evenly balanced that it is difficult or impossible for an IRS auditor to decide whether a worker is an IC or employee, the existence of a written IC agreement can tip the balance in favor of IC status. (See Chapter 12 for guidance on creating your own independent contractor agreement.)

Employee Benefits

Providing a worker with employee-type benefits, such as health insurance, sick leave, paid vacation leave, or pension benefits, is evidence that you intended to create an employment relationship.

IRS W-2 Form

In the eyes of the IRS, filing an IRS W-2 form for a worker shows that you consider the worker to be an employee.

Duration of Relationship

Employees are typically hired for indefinite periods; they continue to work as long as their employers need or want their services. For this reason, if you hire a worker with the expectation that the relationship will continue indefinitely, the IRS believes you probably intended to create an employment relationship.

However, the IRS recognizes that an IC can perform services for the same business for a long time. The relationship between a business and an IC may be long-term in either of two cases:

- The IC has signed a long-term contract.
- The IC's contracts are regularly renewed because the IC does a good job, prices his or her services reasonably, or is readily available to do the work. As long as your contract with an IC is not open-ended, it can last for as long as needed. Just make sure that each contract has a definite end date.

Performing Key Services

Another important factor for the IRS is whether a worker performs services that are key to the hiring firm's regular business. The IRS figures that if the services an IC performs are vital to your regular business, you will be more likely to control how the IC does the job.

For example, a law firm is less likely to supervise and control a painter it hires to paint its offices than a paralegal it hires to work on its regular legal business. However, it is still possible for the paralegal to be classified as an IC. The IRS will examine all the facts and circumstances; this one factor alone does not make a worker an employee. A paralegal hired by a law firm could very well be an IC if, for example, he or she was a specialist hired to help with especially difficult or unusual legal work.

Applying the Test

When you examine your relationship with any worker, you will probably find some facts that support IC status and others that support employee status. This is because ICs are rarely totally unconstrained in performing their contracts, while employees almost always have some degree of autonomy.

You need to weigh all the evidence to determine whether, looking at the relationship as a whole, evidence of control (indicating the worker is an employee) or autonomy (indicating the worker is an IC) predominates. A good way to do this is to draw up a simple chart with factors that indicate IC status on one side and those that show employee status on the other.

The following examples are taken directly from the IRS manual on how to classify workers, which is used to train IRS auditors. They illustrate that the IRS expects to find a mix of facts and that not all factors need to point to IC status for a worker to be classified that way.

> EXAMPLE: An attorney is a sole practitioner who rents office space and pays for the following items: telephone, computer, online legal research linkup, fax machine, and photocopier. The attorney buys office supplies and pays bar dues and membership dues for three professional organizations. The attorney has a part-time receptionist

who also does the bookkeeping. The attorney pays the receptionist, withholds and pays federal and state employment taxes, and files a Form W-2 each year. For the past two years, the attorney has had only one client—a corporation with which there has been a longstanding relationship. The attorney charges the corporation an hourly rate for services and sends monthly bills detailing the work performed for the prior month. The bills include charges for long distance calls, online research time, fax charges, photocopies, mailing costs, and travel costs that the corporation has agreed to reimburse.

Analysis: There are factors here that show control and factors that show autonomy. The attorney has a number of ongoing business expenses that give him a risk of loss—he pays for an office, supplies, professional dues, and an employee receptionist. The fact that he has an employee is very strong evidence of IC status. On the other side of the ledger, he has only one client and is paid by the hour and reimbursed for certain expenses. However, hourly payment and expense reimbursement are common among attorneys, so these factors probably would not be viewed as very important by the IRS. Having a long relationship with a single client points to employee status, but this factor is probably outweighed by the factors showing IC status.

EXAMPLE: A manufacturer's representative is the sole proprietor of a building supplies business and has an exclusive contract with the building supplies manufacturer. The representative has the sole right to the territory covered and sells only that manufacturer's products, but did not pay anything for the right to the territory. The representative has a bachelor's degree in civil engineering and belongs to several professional associations, paying membership dues. The representative has an office and a secretary, but the manufacturer does not reimburse for these expenses. The representative's name appears in the yellow page advertisements under both the representative's sole proprietorship name and the name of the manufacturer represented. The representative is required to provide regular trip reports to the manufacturer and attend sales meetings and trade shows conducted in the representative's territory.

The representative bids on portions of major commercial construction contracts. These jobs require engineering skills and design work to adapt the building materials to the project plans. All bids are subject to the manufacturer's review. Upon winning a bid, the representative engages and pays the workers who will install the building materials, providing the necessary construction bonds. The representative submits invoices to the general contractor for payment directly to the representative on forms prescribed by the manufacturer. If the general contractor fails to pay, the representative is responsible for collecting and is liable to the manufacturer for payment.

Analysis: This complex example has a real mix of factors. However, although several factors point to employee status, it's likely the representative would be classified as an IC because he has such a strong risk of loss. He is personally responsible for hiring and paying the workers who install the manufacturer's building materials. He is also responsible for collecting the amounts due from the building contractor customers. If the contractors fail to pay, the representative must pay the manufacturer out of his own pocket. The representative can easily lose his shirt if he can't collect from a contractor. This enormous financial risk points strongly to IC status and likely outweighs the employee factors.

IRS and Court Rulings for Specific Occupations

This section examines IRS rulings, court decisions, and IRS guidelines that give detailed guidance on how to classify workers in nine occupations:
- architects
- attorneys
- beauty salon and barbershop workers
- limousine drivers
- van operators
- taxi drivers
- television commercial and professional video workers

- truck drivers, and
- companion sitters.

Following these guidelines will give you persuasive ammunition if the IRS ever decides to audit your business and to question how you've classified your workers.

Architects

An architect who performs services for a variety of clients and who assumes professional liability for his or her work product will almost certainly be viewed as an IC by the IRS.

However, the IRS has ruled that an architect could be classified as an employee when circumstantial evidence suggests that the architect is a de facto employee, such as when an architectural firm gives the architect an office, staff support, expense accounts, and instruction on how the project is to be completed. In other words, the firm might be calling the architect an IC, but treating him or her just like an employee.

Therefore, firms using self-employed architects need to be very careful. It's best if the architect performs his or her services from an outside office without any supervision by the firm. The firm's control over the architect should be limited to accepting or rejecting the architect's final work product.

> **TIP**
>
> **Draftspeople are treated differently.** Licensed architects are not the only people who work in the architectural design field. For example, there are also thousands of architectural draftspeople. The IRS generally classifies such workers as employees. These workers are relatively low-skilled and unlicensed. As you may recall, the IRS is more likely to classify low-skill workers as employees because companies generally supervise and direct their work.

Attorneys

A lawyer who offers services to the public and who is hired by a company or individual to litigate or defend a particular lawsuit is an IC. This remains true even when the lawyer spends most or all of his or her time on the case.

And this rule applies even where a lawyer who has a variety of clients is paid an annual retainer to defend any suit brought against the client.

There is a different rule, however, for attorneys who work as associates in law firms. The IRS will deem an attorney to be an employee of a law firm when the firm pays the attorney a salary, provides the attorney with office space and secretarial help, and requires the attorney to work a specified number of hours. This is so even if the firm allows the associate to retain some of the fees the firm earns from some of its cases.

In-house corporate counsel—lawyers who work full-time representing a single company and who do not offer their services to the public—are almost always classified as company employees.

Contract attorneys are lawyers hired by law firms to help out on a per-project basis. Some contract attorneys are hired through employment agencies and others are hired directly. Unlike associates, they are not considered members of the firm. Whether they should be classified as ICs or employees depends on how they are treated on the job.

Beauty Salon and Barbershop Workers

Beauty salons and barbershops are often staffed by ICs. The owner of the shop will typically rent or lease a chair or booth to a barber, beautician, hair stylist, or manicurist—and the IRS will consider all of these people ICs in the right circumstances. When deciding whether to classify these people as ICs or as employees, the IRS is predominantly interested in how they are paid. The IRS has a list of questions that it generally asks the owners of barbershops and beauty shops when it is trying to determine whether a worker is an employee or independent contractor. These questions can be quite instructive for shop owners who want to ensure that they establish IC relationships with their workers:

- How does the worker pay the shop owner for the space he or she uses? Workers who pay a flat fee rather than a percentage of their earnings are usually found to be ICs by the IRS because this shows they have a risk of loss.
- How much is the weekly/monthly rate? The higher the rate, the greater the risk of loss and the more likely the worker is an IC.

- Does the worker rent a particular space? In the case of a barber or hair stylist, renting a particular space—as opposed to simply working anywhere the shopkeeper wants—indicates IC status because it shows that the owner does not control where the worker does his or her work.
- Who is responsible for damage to the chair? If the barber or stylist is responsible, he or she will have a greater risk of loss, which will indicate IC status.
- Who maintains the worker's appointment book? The worker should maintain his or her own book. If the owner maintains the book, it shows control over the worker and indicates employee status.
- Who collects the money earned by the worker? If the worker collects the money he or she earns, this will indicate IC status.
- Who pays for the worker's supplies? Generally, if a worker pays for his or her own supplies, this fact will indicate IC status.
- Who maintains the books and records? If the worker maintains his or her own books and records, this fact will point to IC status.
- How are assistants compensated? ICs pay for their own assistants; shop owners pay for assistants for their employees.

To ensure that the IRS classifies a worker as an IC, the shop owner should not supervise or otherwise control the worker. However, the shop can require the worker to comply with some basic work rules, including the following:

- maintaining a clean work area
- maintaining his or her own tools, and
- providing and maintaining his or her own uniforms.

In most states, barbers and beauticians must have licenses. If they aren't licensed, they will look more like employees to the IRS. The chart below lists some additional factors the IRS may consider.

Factors Showing Beauty Salon Worker Is an Employee	Factors Showing Beauty Salon Worker Is an IC
• Worker is required to wear a uniform. • Worker has required work schedule/hours. • Worker does not handle own sales receipts. • Worker does not make own appointments. • Owner provides training. • Owner provides towels/smocks.	• Worker has key to shop. • Worker makes own schedule. • Worker buys own products. • Worker has separate business phone number.

Source: IRS Publication: *Beauty, Barber, & Cosmetology Industries*; February 2004.

The following examples, taken from IRS Publication 15, *Employer's Tax Guide* (Circular E), illustrate how the IRS views these types of workers.

EXAMPLE 1: Paul is a barber. He signed a lease agreement with Larry, the owner of a barber shop, to use a chair in Larry's shop. Larry bears all the shop expenses, including rent, utilities, advertising, linens, and other supplies. Paul keeps 70% of the receipts from his chair, and Larry keeps 30%. Paul puts all receipts in Larry's cash register. At the end of the week, Larry pays Paul the agreed percentage of the receipts. Paul must comply with the shop hours that Larry has posted on the shop door. Paul must take customers in turn, maintain clean premises, use clean towels and sterile equipment, and keep a clean personal appearance. Larry's income depends on a percentage of Paul's receipts. Larry retains the right to direct and control Paul to protect his investment and to be assured sufficient profit from the shop. Paul has no investment in the shop, assumes no liability for its operation, and furnishes nothing except his personal services. Is Paul Larry's employee?

Yes. Virtually every factor in this example weighs against Paul's being an IC and in favor of his being an employee. Paul has no risk of loss because he has no investment in the shop and because Larry pays all the expenses. The fact that Larry keeps all the receipts in his cash register and pays Paul a percentage at the end of the week also shows that Paul is an employee—having control over all the money gives Larry the right to control Paul's work. Larry's control is also evident by the fact that Paul must work set hours. All these factors are consistent only with a finding of employee status by the IRS.

EXAMPLE 2: Charlie, the owner of a barbershop, and Sally, a professional manicurist, have an agreement under which Sally provides manicuring services to shop patrons during business hours. According to the agreement, Sally regulates her own hours, furnishes her own equipment, and keeps the proceeds from her work. She does not use the shop cash register nor does she report her earnings to Charlie. She sometimes hires a substitute to fill in for her when she doesn't want to work. Charlie cannot direct the way she performs her services. Either of them can end the agreement at any time. Although Charlie has the right to dismiss Sally by ending the agreement, and although he furnishes her a place to work, he does not have the right to direct and control her work. Is Sally an employee?

No. Sally is self-employed. Her paying a fixed monthly fee for the booth, setting her own hours, and selecting her own customers are important factors pointing to IC status. These all show that Charlie doesn't control how Sally works. Sally also has a risk of loss.

Limousine Drivers

The IRS uses a three-prong test to determine whether limousine drivers are employees or ICs. (The word limousine includes sedans, vans, and stretch limousines used in a livery service, but not taxis.) The three determinants are:

- Has the driver made an investment in the limousine, either by owning or leasing it? Drivers who don't make such investments are employees.
- Does the driver have a financial stake in the limousine's profit or loss potential? Drivers without such a risk of loss are employees.
- Does the limousine company have the right to control the driver on the job? If so, the driver is an employee.

RESOURCE

Want more information on drivers? Each of these prongs requires a rather detailed analysis. You can get a complete copy of the IRS guidelines on how to classify limousine drivers from the IRS website at www.irs.gov.

Van Operators

A van operator is the driver of a vehicle that transports household goods, such as furniture and other belongings. The IRS guidelines discussed here cover only those van operators who work under a written agreement with a carrier or agent and who own their own trucks or truck tractors.

To determine whether a van operator is an IC or employee, the IRS examines several different areas of the relationship between the van company and the operator. Those include:

- the operator's financial investment in his or her equipment, either through ownership or lease (this is the most important factor)
- evidence of the operator's potential for profit and loss, including expenses, compensation, and the financial burden of hiring his or her own assistants
- who determines the operator's work schedule and manner of performance, and
- other factors that distinguish an IC from an employee, such as training, independent decision making, and termination issues.

In one of the largest worker misclassification disputes ever, FedEx delivery drivers from 26 states filed a class action lawsuit against the company, claiming they should have been classified as employees instead of ICs. The court examined the laws of 27 states and held that in most cases, the drivers were not employees because FedEx didn't have the right to control how they worked or fire them on a nationwide basis. (*In re FedEx Ground Package System, Inc.*, 2010 U.S. Dist. LEXIS 134959 (N.D. Ind. 2010).)

RESOURCE

Get the IRS rules. You can get a complete copy of the IRS guidelines on how to classify van operators from the IRS website at www.irs.gov.

Taxi Drivers

Some taxi drivers rent company-owned vehicles while others own their own and use the company's dispatch services only. The IRS takes the position that regardless of whether the driver owns or leases the taxi, he or she is an employee of the company if the driver must pay the company a percentage of the fares he or she earns. In order to determine how much money it's owed, the taxi company must conduct an accounting of all the fares collected by the driver during the shift. The IRS asserts that this right to an accounting means that the taxi company has the right to control the driver and that the driver is therefore the company's employee.

On the other hand, taxi drivers who pay a taxi company a fixed amount—rather than a percentage of their earnings—can be ICs, whether they own their taxis or lease them from the taxi company. The taxi company doesn't have to track the fares the driver receives when the driver pays the company a fixed fee.

However, the IRS will classify drivers who pay a fixed fee as employees if other factors indicate that the company controls the driver—for example, if the company requires the driver to accept all company dispatch orders.

In addition, the IRS will classify as employees taxi drivers who work for companies that use a voucher system to bill their customers.

Hollywood Fares

You might get your big break in show biz before you spot a taxi in Los Angeles. Nonetheless, Uncle Sam still wants to know who's really calling the shots between the drivers and dispatchers in L.A. An IRS audit of more than 650 taxi drivers concluded that most were employees, even if they paid the taxi company a fixed fee. The taxi companies required their drivers to follow various rules and regulations, all of which showed that the companies controlled the drivers. For instance, drivers could not refuse dispatch orders from the company without suffering adverse consequences. Also, the company had the right to discharge the drivers at any time. And the cars all had to be painted in the company colors. Finally, all of the drivers were included in the company's property damage insurance policy.

Truck Drivers

Although some trucking companies hire employees to drive company trucks and trailers, most use ICs. The trucking industry uses several different types of ICs, including:

- Owner-operators, who are people who own and operate their own trucks, including tractor-trailers and bobtails.
- Subhaulers, who are people or companies that own and operate a single tractor-trailer or a fleet of tractor-trailers that are then leased to prime carriers. Subhaulers are paid a percentage of the freight bills prepared by the prime carriers.
- Pothaulers, who are a subclass of owner-operators. Pothaulers are owner-operators who pick up full, sealed containers from the harbor and transport them to the terminal of the prime carrier or break-bulk agent. Pothaulers are usually paid a flat rate for each container hauled.

Diapers and Bananas and Taxes, Oh My!

The IRS automatically classifies some truck drivers as employees for employment tax purposes. These include drivers of certain commodities, such as meat and vegetable products, fruit, bakery products, and beverages; they also include drivers of laundry and dry cleaning. (See "Food, Beverage, and Laundry Distributors," below, for a detailed discussion of automatic classification.)

The single most important factor in determining whether a truck driver is an IC or an employee is truck ownership. A trucker who owns or leases his or her own truck may be an IC, but a trucker who uses a company truck will almost always be found to be an employee, even if other factors indicate IC status.

If a truck driver wants to be an IC, it's always best for the driver to buy or lease the truck from someone other than the person or company for whom the driver works. Otherwise, the IRS will analyze the sale or lease very carefully to make sure it isn't a sham designed to help make the driver look like an IC when the driver is really an employee. IRS examiners will review the details carefully—especially the title and sale/lease documents. The driver must pay a commercially reasonable purchase or lease price based on the fair market value of the vehicle and must be personally liable for the payments. If the driver financed the purchase, the driver must pay a reasonable interest rate. If the driver leased the vehicle, the lease should usually be for a minimum of one year. A purchase or lease price is reasonable if it is comparable to prices other companies would charge.

In addition to examining the truck ownership, the IRS will also ask whether the driver:

- pays business and travel expenses
- receives compensation based on a percentage of revenue or miles driven rather than a fixed salary
- hires and pays assistants, drivers, and mechanics
- pays his or her own license fees and road taxes
- maintains his or her own truck storage and maintenance facility or business office
- sets his or her own work schedule
- determines the manner of performing the work
- has a choice to accept or reject jobs
- may delegate services to another driver, and
- works for more than one firm at a time.

Doing the above tasks indicates IC status; not doing them indicates employee status. The following example, taken from IRS Publication 15, *Employer's Tax Guide* (Circular E), illustrates how the IRS classifies truck drivers.

EXAMPLE: A company engages Phil Blue to haul produce to its customers. The company has legal ownership and control of the trucking equipment. The company can require Phil (on an hour's notice) to make deliveries at specific times and specific places. If Phil refuses, he will jeopardize his relationship with the company. He has to operate and maintain the equipment and provide the necessary operators and helpers. He is not allowed to use the company's equipment to haul for others. He is paid on a tonnage basis and is not guaranteed a minimum amount of compensation. He has to pay the operators and helpers out of his tonnage receipts as well as pay for all the insurance coverage required by the company. Is Phil an employee of the company?

Yes, but it's a close call. The fact that Phil is paid on a tonnage basis and not guaranteed any minimum compensation is on the IC side of the ledger. So is the fact that he must personally pay for helpers and all insurance coverage and maintain the vehicles. However, all the other factors indicate employee status. Most importantly, the company owns the truck. The company also exercises substantial control over Phil. He is not allowed to use the company truck to haul for others, and he must accept all assignments offered by the company. The fact that Phil must make deliveries at times and places specified by the company would seem to be a neutral factor—obviously, produce must be delivered according to a certain time schedule.

Companion Sitters

Companion sitters are people who serve as companions for the sick and elderly. Typically, a companion sitter will find employment through a specialized placement service, rather than being hired directly.

A special employment tax law provision provides that sitters are not employees of the placement service if the service does not pay them wages—that is, if the companion sitter is paid directly by the person or business he or she is sitting for.

The relationship of the companion sitter to the client is a little more complicated. To sort it out, use the IRS common law test. Under this test, a sitter would be a client's employee if the client has the right to control how the sitter performs the companion sitting services. It seems likely the right of control would be present in most companion sitting situations, except, perhaps, where the client is so ill or elderly that he or she doesn't have the physical capacity to exercise any control over the sitter.

Companion sitters who perform their services at the client's home are considered household workers for federal payroll tax purposes. (See Chapter 7 for guidance on dealing with household employees.)

Step 3: Check Statutory Employee Rules

This step applies only if you've determined that a worker who is an IC under the common law test discussed in Step 2 (see above) falls into one of the following categories:

- corporate officers
- home workers
- drivers who distribute food products, beverages, or laundry
- full-time life insurance salespeople, and
- traveling or city salespeople.

If a worker you classify as an IC under the common law test comes within one of these categories, and if the additional requirements discussed below are met, the worker is a statutory employee. Statutory employees must be treated as employees for FICA purposes, so you must pay half of their Social Security and Medicare taxes yourself and withhold the other half from their paychecks. You must also pay federal unemployment taxes for statutory employees. You don't have to withhold federal income tax from the workers' paychecks, however (except for corporate officers). The chart below summarizes what you must withhold and pay for statutory employees.

Employer Treatment of Statutory Employees			
Type of Worker	Income Tax Withholding	Social Security Taxes	Federal Unemployment Taxes
Corporate officer	Withhold	Pay and withhold	Pay
Agent or commission driver	No withholding	Pay and withhold	Pay
Life insurance salesperson	No withholding	Pay and withhold	None due
Traveling or city salesperson	No withholding	Pay and withhold	Pay
Home worker	No withholding	Pay and withhold if total pay is more than $100 in cash during the year	None due

You must give every statutory employee a Form W-2, *Wage and Tax Statement,* and check the Statutory Employee designation in Box 15. The W-2 must show the Social Security and Medicare tax withheld, as well as the Social Security and Medicare income. You must also file a copy with the Social Security Administration. You can also download this document from the agency's website at www.irs.gov.

TIP

Don't forget to check the safe harbor. If a worker is a statutory employee, you can still avoid paying employment taxes if you qualify for safe harbor protection, as discussed below. However, most businesses that use statutory employees don't qualify under the safe harbor rules.

There are three threshold requirements that the worker must meet in order to be a statutory employee other than a corporate officer. (If you are dealing with a corporate officer, you can skip ahead to "Corporate Officers," below.)

- The worker must do the work personally. The worker cannot subcontract the work out to others. You can put this requirement in writing, but you don't have to. If you explicitly tell a worker that he or she must perform the services personally, and the worker agrees, the requirement is satisfied. In addition, this requirement can be implied from the circumstances. For example, telling someone: "I want you to clean my house," implies that you want him or her to perform services for you personally.
- The worker must not make a substantial investment in the equipment or facilities used to perform the services. These include things such as office space, furniture, office equipment, and machinery. They don't include tools, instruments, and clothing commonly provided by employees in the trade—for example, uniforms that employees typically provide for themselves or inexpensive hand tools that carpenters typically have. Nor do they include education, experience, or training. Statutory employees can also use their own vehicles for their transportation or for transporting goods.

- The worker must have a continuing relationship with the hiring firm. This includes regular part-time work and regular seasonal employment. But a single project is not enough to constitute a continuing relationship, even if the job takes a long time to complete.

Things to Do If You Don't Want Statutory Employees

You can avoid having a worker classified as a statutory employee by setting up the work relationship so that it does not satisfy any one (or more) of the three threshold requirements described above. For example:

- Sign a written agreement with the worker stating that he or she has the right to subcontract or delegate the work out to others. But the agreement must reflect reality—that is, you must really intend to give the worker the right to delegate.
- Hire workers with a substantial investment in outside facilities.
- Avoid having a continuing relationship with the worker. Use the worker for a single project, not ongoing work. Spread your hiring around by using lots of different workers instead of a few favorite ones. Also, avoid giving workers lengthy projects. Break down complex projects into separate tasks and hire different workers to complete them.

Now that you know the threshold requirements, let's look more closely at the categories of workers eligible to be statutory employees.

Corporate Officers

If a corporation's president, treasurer, vice president, or secretary:

- performs services for the company and
- receives (or is entitled to receive) compensation, the officer is a statutory employee. Compensation doesn't just refer to wages. If the officer earns stock options, then this requirement is met.

EXAMPLE: Murray, his sister Rose, and his brother-in-law Artie start a catering business, which they set up as a corporation. Murray serves as the president, Rose as the secretary-treasurer, and Artie as vice president. Murray and Rose actually run the company. Artie contributed start-up funds, but does not work for the corporation and receives no pay from it. Murray and Rose are statutory employees of the corporation; Artie is not.

Can Safe Harbor Rules Be Used for Corporate Officers?

A corporation that misclassifies corporate officers as ICs may be able to qualify for safe harbor relief under the rules discussed below and avoid having to pay back employment taxes and penalties. To qualify, the corporation must have issued all required Form 1099s to the officers and have had a "reasonable basis" for treating them as ICs instead of employees. Although the IRS has not said so, this safe harbor protection presumably could also be used for other statutory employees, such as home workers and life insurance salespeople.

A reasonable basis could include advice from an attorney or accountant that the officer was an IC or evidence that IC treatment was a longstanding practice in the industry involved (see below for more information on the safe harbor rules). In practice, however, it is difficult for most companies that use statutory employees to show a reasonable basis for classifying them as ICs. In one case, for example, an accountant who incorporated his business was denied the safe harbor. The IRS found that he didn't have a reasonable basis for classifying himself as an IC of the corporation because he performed virtually all of the company's work. (*Spicer Accounting v. United States*, 918 F.2d 90 (9th Cir. 1990).)

Home Workers

People who work in a home (theirs or someone else's) or a workshop are statutory employees if they meet the threshold requirements described above, and if all of the following criteria apply:

- They do the work away from the hiring firm's place of business—usually in their own home or workshop, or in another person's home.

- They do the work on goods or materials furnished by the hiring firm.
- They perform the work in accordance with the hiring firm's specifications; generally, such specifications are simple and consist of patterns or samples.
- The hiring firm requires the worker to return the processed material to the hiring firm or to some person designated by the hiring firm.

This group usually includes people who make or sew buttons, quilts, gloves, bedspreads, clothing, needlecraft products, and similar goods. However, it can include other workers as well.

EXAMPLE: An educational consultant who collected data on schools and collated the data at home using a template provided by the hiring firm qualified as a home worker for statutory employee purposes. The Tax Court reasoned that all the requirements for home worker status described above were satisfied because:
- the consultant did the work away from the hiring firm's premises
- the template provided to the consultant was a "material" within the meaning of the rule
- the consultant was given specific instructions on which schools to visit, the type of data to be collected, and the format for presenting it, and
- the consultant performed services on materials furnished by the hiring firm and was required to return the materials to it. (*Van Zant v. Comm'r*, TC 2007-195.)

If all these requirements are met, you must pay the employer's share of FICA. However, you don't have to pay FICA tax if you pay the home worker less than $100 for a calendar year.

If the worker doesn't qualify as an employee under the IRS common law test, you don't have to withhold payroll taxes or pay federal unemployment taxes (FUTA).

EXAMPLE: Rosa sews buttons on shirts and dresses. She works at home. She does work for various companies, including Upscale Fashions, Inc. Upscale provides Rosa with all the clothing and the buttons she must sew. The only equipment Rosa provides is a needle. Upscale gives

Rosa a sample of each outfit showing where the buttons are supposed to go. When Rosa finishes each batch of clothing, she returns it to Upscale. Rosa is a statutory employee. Upscale must pay employer FICA (and withhold Rosa's share of FICA). But it need not pay FUTA or withhold federal income tax unless Rosa is an employee under the common law test.

Are Technological Home Workers Statutory Employees?

The home worker statutory employee rules are over 60 years old. They were meant to apply primarily to people in the garment trade who worked at home on a piecework basis and whose only investments were needles and thread. Today, a vast array of professionals work at home at technologically oriented occupations undreamed of years ago, such as computer programming, website design, video editing, and many others. Should these home workers be classified as statutory employees? Usually, no. Such professionals should not be so classified because:

- they usually have a substantial investment in equipment, such as home computers, and
- they generally aren't furnished materials or goods by the hiring firm.

Food, Beverage, and Laundry Distributors

If they meet the threshold requirements discussed above, drivers who distribute meat or meat products, vegetables or vegetable products, fruits or fruit products, bakery products, beverages other than milk, or laundry or dry cleaning are statutory employees.

The products these workers distribute may be sold at retail or wholesale. The drivers may either be paid a salary or by commission. They may operate their own trucks or trucks belonging to the hiring firm. Ordinarily, they service customers designated by the hiring firm as well as those they solicit.

EXAMPLE: Alder Laundry & Dry Cleaning enters into an agreement with Sharon to pick up and deliver clothing for its customers. Sharon has similar arrangements with several other laundries and arranges her route to serve all the laundries. None of the companies has any control over how she performs her services. She owns her own truck and is paid by commission. Sharon qualifies as an IC under the common law test.

However, she is a statutory employee because all three threshold requirements are met: Her agreement with Alder acknowledges that she will do the work personally, she has no substantial investment in facilities (her truck doesn't count because it's used to deliver the product), and she has a continuing relationship with Alder. Alder must pay employment taxes for Sharon—that is, pay half of her FICA and withhold the other half from her pay, and pay all of the applicable FUTA. However, Alder need not withhold federal income tax because Sharon is an IC under the common law test.

Delivery people who buy and sell merchandise on their own account or deliver to the general public as part of an independent business do not fall within this category. It's easy to tell whether drivers have independent businesses. Instead of merely delivering products owned and sold by others, they buy the products themselves and resell them.

Life Insurance Salespeople

This group of statutory employees includes salespeople whose full-time occupation is soliciting life insurance applications or annuity contracts, primarily for one life insurance company. The company usually provides work necessities, such as office space, secretarial help, forms, rate books, and advertising material.

If a life insurance salesperson is a statutory employee, the hiring firm must pay FICA taxes. However, it need not pay FUTA taxes or withhold federal income taxes unless the salesperson is an employee under the common law test.

EXAMPLE: Walter Neff sells life insurance full time for the Old Reliable Life Insurance Company. He works out of Old Reliable's Omaha office, and the company provides him with a desk, clerical help, rate books, and insurance applications. Walter is a statutory employee of Old Reliable. The company must pay employer FICA and withhold employee FICA from Walter's pay. It need not pay FUTA or withhold federal income tax unless Walter is an employee under the common law test.

Business-to-Business Salespeople

Salespeople are statutory employees if they meet the threshold requirements described above and they:

- work full time for one person or company, except, possibly, for sideline sales on behalf of someone else
- sell on behalf of, or turn their orders over to, the hiring firm
- sell merchandise for resale or supplies for use in the buyer's business operations (as opposed to goods purchased for personal consumption at home), and
- sell only to wholesalers, retailers, contractors, or those who operate hotels, restaurants, or similar establishments; this does not include manufacturers, schools, hospitals, churches, municipalities, or state and federal governments.

This group doesn't include drivers who distribute food, beverages, or laundry. (For more on these drivers, see "Food, Beverage, and Laundry Distributors," above.)

Generally, this category includes traveling salespeople who might otherwise be considered ICs. Such salespeople are ordinarily paid on a commission basis. The details of their work and the means by which they cover their territories are not typically dictated to them by others. However, they are expected to work their territories with some regularity, take purchase orders, and send them to the hiring firm for delivery to the purchaser.

EXAMPLE: Linda sells books to retail bookstores for the Simply Sons Publishing Company. Her territory covers the entire Midwest. She works only for Simply and is paid a commission based on the amount of each sale. She turns her orders over to Simply, which ships the books to each bookstore customer. Linda is Simply's statutory employee. The company must pay FICA and FUTA taxes for her. However, Simply need not withhold federal income taxes from her pay unless she qualifies as an employee under the common law test.

Step 4: Check the Safe Harbor Rules

The employer's safe harbor is a set of legal protections found in a footnote to Section 530 of the Internal Revenue Code. It is called safe harbor protection because it provides employers with a refuge from the cold winds and turbulent waters of an IRS audit. If you meet the three requirements for safe harbor protection discussed below, the IRS can't impose assessments or penalties against you for worker misclassification, and you can safely and confidently treat the workers involved as ICs.

The safe harbor was intended to help employers who often have a difficult time determining how to classify their workers under the IRS common law test. If the safe harbor requirements are met, an employer may treat a worker as an IC for payroll tax purposes even if the worker should have been classified as an employee under the common law test discussed in Step 2, above.

Unfortunately, experience has shown that few employers are able to take advantage of the safe harbor. Most can't satisfy the three requirements discussed in this section. However, this doesn't mean you can't satisfy them. If you think a worker should be classified as an employee under the common law test, or you're not sure how to classify a worker under the test, look at the safe harbor rules carefully. If you can meet them all, you can treat the worker as an IC for payroll tax purposes.

To receive safe harbor protection for a worker, you must do each of the following:

- file all required 1099-MISC forms for the worker
- consistently treat the worker—and others doing substantially similar work as that worker—as an IC, and
- have a reasonable basis for treating the worker as an IC.

If you own more than one company—for example, a parent corporation that has one or more subsidiary corporations—the safe harbor rules apply to each company separately. If workers for one company don't qualify for the safe harbor, those who work for a different company owned by the same person might. Likewise, the fact that workers for one company do qualify for safe harbor protection doesn't mean that those who work for other companies qualify. (IRS Field Service Advice 200129008 (December 15, 2000).)

CAUTION

Safe harbor applies only to federal employment taxes. It is very important to understand that the Section 530 safe harbor doesn't make workers ICs, validate them as ICs, or convert them from employees to ICs. Rather, Section 530 classifies the affected workers as "nonemployees" for federal employment tax purposes only—that is, for purposes of Social Security and Medicare taxes. It has no application to federal or state income taxes, workers' compensation, state unemployment taxes, or pension plan rules. Thus, if a worker qualifies for the safe harbor, you can safely treat him or her as an IC only for federal employment tax purposes. You must apply the normal rules to determine whether the worker is an IC or employee for all other purposes. For example, if the worker doesn't qualify as an IC under the regular IRS common law test, you'll have to withhold income taxes from the worker's pay and treat the worker as an employee for purposes of Obamacare, employee pension and profit-sharing plans, and federal labor laws. You might also have to provide unemployment insurance and workers' compensation for the worker, depending on your state's tests.

First Prong: Filing IRS Form 1099-MISC When Appropriate

The first requirement you have to meet to obtain safe harbor protection is to file IRS Form 1099-MISC for any unincorporated IC to whom you paid $600 or more in any year after 1977. You must file this form with the IRS by February 28 of the year after the year in which the individual performed work for you.

This simple requirement is a big stumbling block for many businesses because they fail to file 1099s for workers they treat as ICs. Some are unaware of the requirement, while others fear that filing 1099s may get them audited.

Better Late Than Never

The IRS has interpreted the 1099 requirement very strictly, ruling a business must file all required 1099s on time—that is, by February 28 of the year following the year the IC worked for the firm (see Chapter 11)—to be eligible for safe harbor protection. However, in an important victory for hiring firms, the U.S. Tax Court has held that the safe harbor is available as long as a firm files all required 1099s sometime; they don't have to be filed exactly on time. In this case, a firm that hired emergency room doctors filed its 1099s ten weeks late. The IRS said it could not use the safe harbor because of its tardy filing, even though it satisfied the other requirements. The Tax Court said the firm should be given safe harbor relief. As a result, the firm didn't have to pay over $250,000 in back taxes and penalties. (*Medical Emergency Care Assocs., S.C. v. Comm'r*, 120 T.C. 426 (2003).)

If you've failed to file 1099s for workers who otherwise qualify for safe harbor protection, you should do so now, even though the filing is late. And if you're audited, you should insist that the IRS follow the *Medical Emergency Care* case if the only thing standing between you and the safe harbor is your failure to timely file 1099s.

You need not file Form 1099-MISC for the following types of ICs:
- ICs to whom you paid less than $600 during the year
- incorporated ICs (except for incorporated doctors and lawyers), and
- ICs who performed services not related to your business (for example, household workers).

If a Form 1099-MISC is not required, you can obtain safe harbor protection without filing it.

It's up to you to prove that you filed the 1099s, so be sure to keep copies and some proof of the date of postal or electronic mailing. If you lack proof, the IRS can check its records to see whether the forms were filed, but this requires a time-consuming computer run.

Second Prong: Consistent Treatment

To qualify for the safe harbor, you must treat all workers who hold substantially similar positions as independent contractors for federal tax purposes. If you treat even one of the workers as an employee for federal tax purposes, then you cannot use the safe harbor rules for any workers who hold substantially similar positions, even if you've always treated those workers as ICs.

How do you treat a worker as an employee for federal tax purposes under this prong? You do any one of the following:

- withhold federal income, Social Security, or Medicare taxes from the worker's wages, whether or not you actually pay the withheld money to the IRS
- file a federal employment tax return for the worker (IRS Forms 940 through 943), or
- file a W-2 *Wage and Tax Statement* for the worker, whether or not you actually withhold the tax.

An IRS auditor can easily discover whether you've consistently treated workers performing similar services as ICs by examining your payroll tax returns and similar records.

> EXAMPLE: A roofing company hired 57 people to work as roofing applicators. The company consistently treated 56 of the applicators as ICs. It didn't withhold payroll taxes, file employment tax returns, or provide the workers with W-2 statements. However, the company filed IRS Form 941, the *Employer's Quarterly Federal Tax Return,* for one applicator and paid his Social Security and Medicare taxes. The IRS ruled that the company could not use the safe harbor rules for any of the applicators because they were not all treated as ICs for federal tax purposes.

The consistent treatment requirement applies only to how you treat your workers for *federal tax purposes*. You can treat workers as employees for other purposes and still obtain safe harbor protection. (See below for more about this issue.)

The consistent treatment requirement applies only to workers who hold substantially similar positions. Workers in different positions can be treated differently.

> **EXAMPLE:** The Reliable Building Co. hired ten painters and ten bricklayers in one year. It treated all the painters as employees—that is, it withheld payroll taxes from their pay, issued them W-2 forms, and filed employment tax returns for them. It treated all of the bricklayers as ICs, filing 1099-MISC forms for them. Reliable satisfied the consistent treatment prong of the safe harbor rules as to the bricklayers. The fact that Reliable treated the painters as employees doesn't matter because the painters didn't hold positions that were substantially similar to the bricklayers—that is, being a bricklayer is not substantially similar to being a painter.

The consistent treatment requirement for Section 530 safe harbor relief has one minor exception. The requirement does not apply to services performed after December 31, 2006 by an individual as a test proctor or room supervisor assisting in the administration of college entrance or placement examinations, provided that:

- such services are performed for a tax-exempt nonprofit organization, and
- the organization does not pay employment taxes for the person.

Substantially Similar Positions

How do you know whether workers are in substantially similar positions? The IRS looks at the day-to-day services the workers perform and the method by which they perform them—for example:

- A beauty parlor owner classified cosmetologists who leased their chairs from the owner as ICs and those who worked directly under the owner's control as employees. The owner couldn't qualify for the safe harbor because the cosmetologists held substantially similar positions. (*Ren-Lyn Corp. v. U.S.*, 968 F.Supp. 363 (N.D. Ohio 1997).)

- A sign company classified salespeople to whom it paid a salary as employees and those who worked for commissions on an as-needed basis as ICs. The positions were substantially similar, so no safe harbor. (*Lowen Corp. v. U.S.*, 785 F.Supp. 913 (D. Kan. 1992).)
- A shuttle company had two sets of drivers: those who worked under the company's contract with Illinois Bell, who were treated as employees, and those who worked under a contract with Alamo, who were treated as ICs. The positions were substantially similar, so no safe harbor. (*Leb's Enterprises v. U.S.*, 11385 AFTR2d 2000-886 (N.D. Ill. 2000).)

Thus, to preserve your right to safe harbor protection, you must keep the work that employees and ICs do separate—that is, ICs and employees can't perform similar functions. For example, a trucking company can't classify some drivers as ICs and some as employees and qualify the drivers for the safe harbor. The company would have to classify all of the drivers as ICs to satisfy the consistency requirement. However, the company could classify a bookkeeper as an employee without jeopardizing the safe harbor for the IC drivers.

Before you hire an employee, make sure you haven't used ICs to perform similar services. And before you hire an IC, make sure you haven't used employees to perform similar services.

Timing Concerns

The moment you treat any worker as an employee, you lose safe harbor protection for all workers performing substantially similar services. But you are not prevented from obtaining safe harbor protection for prior years—that is, you can still qualify for the safe harbor for the years before you treated any worker in a substantially similar position as an employee.

EXAMPLE: Quickie Roofing treats all its applicators as ICs in 2012 and 2013. It does not withhold federal taxes, file employment tax returns, or give the workers W-2 forms. In 2014, however, Quickie begins treating some applicators as employees—that is, it does withhold federal tax from their paychecks and files employment tax returns for the workers. Quickie may still obtain safe harbor protection for the applicators for 2012 and 2013, but not 2014 or afterwards.

If you've purchased your company from somebody else, the prior owner's classification practices apply to you as well as the prior owner. Thus, you can't qualify for the safe harbor if a prior owner treated as employees workers performing substantially similar services as workers you wish to classify as ICs.

EXAMPLE: Joe started Acme Trucking in 2010, treating all of his truckers as employees. In 2014, Joe sold the company to Eve. She treats all the truckers as ICs. If the IRS audits Acme for the years after the sale, Eve cannot obtain safe harbor protection for the truckers because Joe did not treat them as ICs.

This rule also prevents hiring firms from evading the consistency requirement by reincorporating their businesses.

EXAMPLE: Ace Trucking, Inc. treats all its drivers as employees. Ace's owners want to convert all the drivers to IC status. The owners dissolve the corporation and transfer all the assets to a new corporation, Joker Trucking, Inc. Joker hires all of Ace's old drivers and classifies them as ICs. Joker Trucking's drivers do not qualify for the safe harbor.

Inconsistent Treatment for Other Purposes

As stated above, the consistent treatment requirement means you must treat your workers consistently for federal tax purposes. You can treat workers differently for other purposes and still obtain safe harbor protection. For example, you could provide a worker with workers' compensation coverage or pay state unemployment compensation taxes for him or her—both of which mean that you are treating the worker as an employee for state tax purposes—and still obtain safe harbor protection if you meet the other prongs.

You may wish to do this if your state has particularly strict rules about which workers qualify as ICs for unemployment insurance or workers' compensation purposes. (See Chapter 5 for information about state taxes, including unemployment insurance. See Chapter 6 for information about workers' compensation.)

Don't File Federal Tax Forms for Independent Contractors

If you classify a worker as an employee for state law purposes, don't use any federal tax forms to report your state tax payments or withholding for the worker. If you file such a form, the IRS will think that you are treating the worker as an employee for federal tax purposes—and you'll lose your safe harbor protection for that worker and for all workers who hold substantially similar positions.

For example, if you pay state unemployment taxes for a worker you classify as an IC for federal tax purposes, don't include the payments in IRS Form 940, *Employer's Annual Federal Unemployment Tax Return*. List on Form 940 only the state unemployment compensation taxes you've paid for workers you classify as employees for IRS purposes as well as state purposes.

Workers With Dual Status

It's possible to treat the same worker as both an IC and employee without violating the consistent treatment prong. This can occur where a worker performs very different services for the same company. For example, an employee bookkeeper for a business might be hired as an IC to design and print an advertising brochure. The fact that the bookkeeper is treated as an employee for the bookkeeping services does not prevent the hiring firm from obtaining safe harbor protection for the worker for the design and printing services.

Third Prong: Having a Reasonable Basis for IC Classification

The final and usually most difficult requirement you must satisfy to obtain safe harbor protection is showing that you had a reasonable basis for treating the workers as ICs. Reasonable basis is another way of saying that you had a good reason for the IC classification.

There are several ways you can show a reasonable basis for treating a worker as an IC; some are simple, some are more complex. These include showing that:

- It is a long-standing practice in your trade or industry to treat similar workers as ICs.
- You relied on advice from an attorney or an accountant that the worker was an IC.
- You relied on past court decisions or IRS rulings.
- You relied on a past IRS audit.
- You had some other good reason for classifying the worker as an IC, such as following a favorable ruling from a state agency.

Ideally, you should take whatever steps are necessary to establish a reasonable basis before you start treating a worker as an IC. If you have already hired ICs, don't wait until you're facing an audit to find a reasonable basis. Find one right now and keep your documentation on file in case of an audit. If you're already undergoing an audit, you can still establish a reasonable basis, but your task may be harder.

Let's look in more detail at each of the ways to prove reasonable basis.

Industry Practice

One way to establish a reasonable basis for treating a worker as an IC is showing that businesses in your geographic area treat similar workers as ICs. In other words, if everybody else does it, you can do it, too. However, in practice, it can be difficult to prove this to the IRS or a court.

To satisfy this requirement, you must prove that you relied on the fact that a significant number of firms in your geographic area use the same classification as a long-standing practice. Let's break this down a bit.

Reliance. If audited, it's acceptable for you to say that you "just knew" everyone else used the same classification. But it's better if you can point to some hard evidence, like a survey of your industry or sworn statements from other business owners. In addition, it's good to have a paper trail showing that you classified the workers as ICs because it was industry practice—for example, corporate minutes stating that the workers involved are being classified as ICs because this is a long-standing practice in your industry. If you don't have anything in writing to show your reliance on the long-standing practice, all is not lost. If you are ever audited, the auditor will interview you and other key people in your business and ask you why you classified the workers involved as ICs. This will give you the opportunity to explain that you relied upon the long-standing industry practice. The auditor will also interview the workers themselves to determine what reasons you gave them for classifying them as ICs, so make sure you tell them that their IC classification was based on industry practice. It's also a good idea to state this in your written IC agreements with the workers. (See Chapter 12 for a discussion of independent contractor agreements.)

Significant number. How many are enough? If you can show that 24% of the businesses in your area treat similar workers as ICs, you'll meet this part of the test. A smaller number may satisfy the IRS, depending on the circumstances.

Geographic area. You don't have to prove that everybody in the world treats workers similar to yours as ICs. You only need to look at the geographic area where you do business. If the market in which you compete is limited to your city, then you need only show that firms in your city do the same thing. If, however, you compete in regional or national markets, the geographic area may include that whole region or even the entire United States.

Long-standing practice. No fixed length of time is required, but the IRS will presume that any practice that has existed for ten years or more is long standing. If you're in a new business or industry—for example, biotechnology—a much shorter time period should be acceptable.

Relying on Professional Advice

For many businesses, the easiest way to establish a reasonable basis is to ask an attorney or accountant whether the workers involved qualify as ICs for IRS purposes. If the answer is "Yes," your reliance on this advice constitutes a reasonable basis. However, this will work only if you obtain the advice before you begin treating the workers involved as ICs.

> **EXAMPLE:** In 1983, the president of a Tennessee company that hired numerous telemarketers to sell gourmet foods contacted his longtime CPA and asked him whether the workers qualified as ICs. The CPA, who was thoroughly familiar with the company's business, advised the owner that the telemarketers qualified as ICs under the IRS common law test. The company classified all of the telemarketers as ICs for federal tax purposes, and filed all required Form 1099s for them. The IRS audited the company in 1991 and claimed that the telemarketers should have been classified as employees during 1989–1990. The company appealed and won. The court held that it was reasonable for the company to rely on advice from its CPA that the workers were ICs. This reliance satisfied the reasonable basis requirement for safe harbor protection. Because the company consistently treated all the telemarketers as ICs, filed all required Form 1099s, and satisfied the reasonable basis requirement, it qualified for safe harbor protection. As a result, the company didn't have to pay a $3,888,918 IRS assessment. (*Smoky Mountain Secrets, Inc. v. United States*, 910 F.Supp. 1316 (E.D. Tenn. 1995).)

Get advice about worker status in writing: A letter from the attorney or accountant will do. The letter should explain why the workers qualify as ICs. Keep the letter in your IC file. Once you obtain such a letter, you have the assurance of knowing you'll probably qualify for safe harbor protection as long as you satisfy the other two requirements—filing all Form 1099s and consistently treating the workers as ICs. If you're audited, you won't have to go through the grueling and uncertain process of proving that the workers involved really are ICs under the common law test.

The IRS says you can't rely on advice from just any attorney or accountant; you must go to one who is familiar with business tax issues. You are not required to go to a high-priced tax specialist, nor do you have to independently investigate the attorney or accountant's credentials. However, you must make sure the person advises businesses. For example, the IRS says you couldn't reasonably rely on the advice of a patent attorney because such attorneys ordinarily don't advise businesses on tax matters.

Relying on Court Decisions and IRS Rulings

If any federal court—that is, federal tax court, federal district court, federal court of appeals, or the U.S. Supreme Court—holds that workers in similar situations were not employees for federal tax purposes, you can use that holding as a reasonable basis for the safe harbor rules.

Published IRS revenue rulings may also be used to show a reasonable basis. You may not rely on IRS private letter rulings or technical advice given to other taxpayers. However, you may rely on advice the IRS gives specifically to you. This includes a specific letter ruling or determination letter from the IRS. If you've purchased your business from someone else, you may not rely on a letter ruling issued to your predecessor.

All you need is one supportive opinion or ruling. The ruling need not be from a federal court in your state or judicial district. It doesn't matter if other opinions or rulings go against you, or if your business is outside of the court's jurisdiction. However, the decision or IRS ruling must have been in existence and not overruled at the time you first began treating the workers involved as ICs. Moreover, you should be able to show that you relied upon the ruling at the time you first treated the worker as an IC.

The facts involved in such court decisions or IRS rulings do not have to be absolutely identical to your situation. Nor do they necessarily have to involve the particular industry or business in which you're engaged. They just have to be similar enough for you to rely upon them in good faith. You should provide the IRS auditor with a copy of the court decision or ruling you relied upon and explain why it was reasonable for you to do so.

Most of the time, you probably won't be able to find a prior court decision or IRS ruling upon which to rely. There are thousands of published IRS rulings, but the vast majority hold that the workers involved were employees. There are not as many federal court rulings on this issue, and the chances of finding a favorable one that may be reasonably relied upon are fairly slim. However, there are favorable court rulings for some industries—for example, cases involving drywall installers, construction workers, and truckers who own their own trucks.

If no other reasonable basis is available, do some legal research to try to find a favorable court decision or IRS ruling before you classify a worker as an IC. You can do the research yourself, or hire an attorney or CPA to do it for you. (See Chapter 13 for information on doing your own research.) If you find a favorable ruling, be sure to carefully document that you relied upon it when you first treated the worker as an IC.

Relying on Past IRS Audits

If your business was audited by the IRS any time after 1977, you could be in luck: The audit could establish a reasonable basis for treating your workers as ICs. If, after the audit, you were not assessed back employment taxes for workers you classified as ICs, the IRS cannot later claim that any of your workers holding substantially similar positions are employees.

The audit must have been for your business. An audit of your own personal tax returns doesn't count, nor do audits of your workers' taxes. And you can't rely on audits by state and local tax authorities. If you own more than one company, the IRS won't allow you to rely on a prior audit of a company other than the one involved in the current audit. And a past IRS audit won't help provide safe harbor relief if you committed fraud.

If you were audited before January 1, 1997, you can use that audit for the safe harbor rules even if it didn't focus on worker classification or employment taxes. Things aren't so simple for audits that occurred after that date, however. A post-1996 audit qualifies as a reasonable basis for purposes of the safe harbor rules only if it included an examination of whether the individual workers involved in the current audit or others holding substantially similar positions were properly classified as ICs for employment tax purposes.

EXAMPLE: The IRS examined Acme Corporation's 1996 income tax return in 1997. The auditor did not ask or consider whether ICs hired by Acme qualified as such for employment tax purposes. The IRS audits Acme again in 2014, and this time the auditor questions the workers' IC classification. Acme may not use the 1997 audit as a reasonable basis because the audit did not include an examination of the worker classification issue.

The prior audit is useful to you only if the workers from that audit (that is, the workers whose IC status was not questioned by the IRS) hold substantially similar positions to the workers who are currently at issue.

EXAMPLE: A nursery and landscaping company hired landscapers and treated them as ICs. The IRS audited the business and determined that the landscapers were ICs. The corporation later hired janitorial workers and also treated them as ICs. The IRS audited the company again and claimed the janitorial workers were employees. The corporation claimed that the prior audit was a safe harbor. The IRS refused to grant safe harbor protection because the janitors and landscapers did not hold substantially similar positions. (*Lambert's Nursery and Landscaping, Inc. v. U.S.*, 984 F.2d 154 (5th Cir. 1990).)

The safe harbor rules do not define "substantially similar." The IRS takes the narrow view that it means that the workers must do the same work. However, some courts take a much more liberal approach. They compare the structure of the relationship between the workers and hiring firm, not the type of work involved. This means that workers in different industries could be considered substantially similar for safe harbor purposes.

EXAMPLE: The landscaping company in the above example appealed its case to federal court and won. The court held that the landscape workers and janitors held substantially similar positions for safe harbor purposes. Even though they did different kinds of work, their relationship with the company was very similar. Both groups of workers were paid on a per job basis and were treated similarly in terms of control, supervision, and work demands. Because the prior audit did not challenge the IC status of the landscape workers, the court ruled that the hiring firm qualified for safe harbor protection for its subsequently hired janitors.

An Audit by Any Other Name Is Not an Audit

Not all IRS contacts are considered audits. An audit occurs only when the IRS inspects your business's books and records. Mere inquiries or correspondence from an IRS Service Center do not count.

For example, a letter you receive from an IRS Service Center to verify a discrepancy in a Form 1099 you filed for a worker is not an audit. The IRS calls such contacts adjustments. If, however, IRS correspondence includes an examination or inspection of your records to determine the accuracy of deductions claimed on your tax returns, then you've been audited.

Compliance checks in which the IRS asks if a business has filed all required returns, including employment tax returns, do not constitute audits. You'll receive a letter from the IRS saying as much if it conducts a compliance check of your business. However, a compliance check would qualify as an audit if the IRS asks about the reason for your worker classification practices or examines books and records other than those forms the IRS requires you to maintain, such as your business's income and employment tax returns.

Worker classification rulings the IRS makes in response to the filing of Form SS-8 by workers or hiring firms also do not constitute audits for Section 530 purposes.

Be sure to keep all copies of your correspondence and other documentation from the IRS as well as the auditor's business card. Show the auditor copies of this documentation to establish you had a prior audit. The auditor will then verify this by checking IRS records.

Catchall Provision

If you can't find a reasonable basis in any of the ways discussed above, you might still be able to squeak through using the catchall provision. If you can demonstrate in any other way a reasonable basis for treating the workers as ICs, you still have a chance at that safe harbor. There are many possible ways to show such a reasonable basis, for example:

- **Determinations by government agencies.** A determination by a state agency or federal agency other than the IRS that the workers involved qualify as ICs may constitute a reasonable basis if the state or federal agency uses the same common law test as the IRS and interprets it similarly. Many states use the common law test for unemployment compensation purposes (see Chapter 5 for a list of these states), and most states use the common law test for workers' compensation purposes (see Chapter 6 for a list of these states). Decisions by federal agencies, such as the U.S. Department of Labor, could also provide a reasonable basis.

- **Good-faith application of common law test.** Even more significantly, courts have held that a reasonable basis can include a hiring firm's reasonable, good-faith—though possibly mistaken—belief that the workers qualified as ICs under the IRS common law test. (See Step 2, above, for a discussion of the common law test.)

EXAMPLE: A firm that provided specialized temporary registered nurses to hospitals classified the nurses as ICs. The firm thought the classification was correct under the common law test. The IRS disagreed and claimed that the firm should have classified the nurses as employees. The firm appealed to federal court. The court held that it could use the safe harbor because it had reasonably concluded that the nurses qualified as ICs under the IRS common law test. This was enough to provide a reasonable basis for the classification. (*Critical Care Registered Nursery, Inc. v. U.S.*, 776 F.Supp. 1025 (D.C. Pa. 1991).)

Reason Is in the Eye of the Beholder—And the IRS

You don't have to use one of the reasons discussed in this chapter to establish reasonable basis. You are limited only by your imagination—and the IRS.

Firms have successfully argued reasonable basis in the following situations:

- A firm relied upon information from an industry or trade association or similar group that the workers could be classified as ICs.
- A firm relied upon a favorable court decision or IRS ruling issued after it began treating the workers involved as ICs.
- A firm relied upon an IRS private letter ruling or determination letter issued to one of its competitors stating that similar workers qualified as ICs.

The IRS has stated that the following reasons do not serve as a reasonable basis for classifying workers as ICs:

- The workers didn't have Social Security numbers.
- It's cheaper for a hiring firm to treat the workers as ICs because it doesn't have to pay half their Social Security taxes or provide them with benefits.
- The workers asked to be treated as ICs so they wouldn't have to have taxes withheld from their pay.

Limitations on Safe Harbor Protection

Even if you meet the three-prong test described above, you will not be allowed to use safe harbor protection if you are a broker who deals with technical service workers or if you want the protection to shield you during an audit of your pension plan.

Technical Service Firms

If you are a broker who contracts to provide certain technical service workers to hiring firms, you may not claim the workers are ICs using the safe harbor rules (though the hiring firm may).

Such brokers—also called technical services firms or consulting firms—may not use the safe harbor if they contract to provide third-party clients with:

- engineers
- designers
- drafters
- computer programmers
- systems analysts, or
- other similarly skilled workers.

EXAMPLE: Acme Technical Services is a broker that provides computer programmers to others. Acme contracts with Burt, a freelance programmer, to perform programming services for the Old Reliable Insurance Company. Reliable pays Acme, which in turn pays Burt after deducting a broker's fee. Acme is a broker, and Old Reliable is a hiring firm. Acme cannot claim that Burt is an IC, but Old Reliable may be able to.

Retirement Plan Audits

The IRS has a special group of auditors called the Employment Plans and Exempt Organizations Divisions, or EP/EO, who audit retirement plans. The IRS uses the common law test to determine whether workers are employees or ICs for retirement plan purposes. Safe harbor protection is unavailable in such audits. If the IRS determines that workers you've classified as ICs are really employees for purposes of your retirement plan, the plan could lose its tax-qualified status, resulting in substantial tax liability.

TIP
If you have a retirement plan, don't rely solely on the safe harbor. If you are audited by the IRS and win your audit through the safe harbor rules, the IRS can still refer your case to the EP/EO for audit. For this reason, if your company has a tax-qualified retirement plan, you should make sure workers you've classified as ICs qualify as such under the IRS common law test in addition to the safe harbor.

IRS Audits

Why Audits Occur 101

Audit Basics 102

    Who Gets Audited 102

    Audit Time Limits 103

    What the Auditor Does 104

    What You Need to Prove in an Audit 105

    Audit Results 107

    Appealing the Audit 107

The Classification Settlement Program 109

    When the CSP Is Used 109

    CSP Offers 110

    CSP Procedures 112

Voluntary Classification Settlement Program 113

    Who Qualifies for the VCSP? 113

    How to Apply 113

    What Happens If You're Accepted? 114

IRS Assessments for Worker Misclassification 116

    Penalties for Unintentional Misclassification 117

    Penalties for Intentional Misclassification 120

Penalties for Worker Misclassification 122

    Trust Fund Recovery Penalty 123

    Other Penalties 124

Interest Assessments 125

Criminal Sanctions 125

Obamacare Penalties 126

Retirement Plan Audits 126

    Tax-Qualified Retirement Plans 126

    Antidiscrimination Rules 127

    Losing Tax-Qualified Status 127

Worker Lawsuits for Pensions and Other Benefits 128

T his chapter provides an overview of IRS audits and explains the considerable assessments and penalties that may be imposed on companies that misclassify workers. It also describes IRS initiatives: the Classification Settlement Program and Voluntary Classification Settlement Program, which allow many businesses to pay reduced assessments if they agree to reclassify the workers involved as employees.

RESOURCE
Want to know more about audits? Dealing with IRS audits is a complex subject. This chapter does not cover the entire audit process in detail. For detailed information on handling IRS audits, see *Tax Savvy for Small Business*, by Frederick W. Daily (Nolo).

Why Audits Occur

Worker misclassification by employers is one of the IRS's top audit priorities. The agency believes that millions of employers routinely misclassify employees as ICs to save on employment taxes. The IRS claims this costs the U.S. government billions in taxes every year.

Thus, it's safe to say that whenever you classify a worker as an IC, you become a potential IRS target. The IRS would prefer all workers to be classified as employees, not ICs. That way, it could collect workers' income and Social Security taxes directly from their employers through payroll withholding.

If you stay in business long enough, and if you have ICs, you'll probably be audited at least once by the IRS. Some businesses are audited far more often. Repeat audits are especially likely if past audits have turned up serious problems. You should always be prepared to defend your worker classification practices to the IRS. At the very least, you want to put yourself in a position where the IRS can impose only the minimum assessments and penalties allowed by law if it determines you have misclassified employees as ICs.

Audit Basics

An audit is an IRS examination of your business, its tax returns, and the records used to create the returns.

The IRS can audit your business for many reasons—for example, to determine whether:

- your business has paid the correct employment taxes for employees and filed the proper forms
- you've withheld the proper income taxes from employees' pay, or
- your employee pension plan meets the federal requirements for tax qualified status. (See "Retirement Plan Audits," below.)

Who Gets Audited

There are a number of ways you can be chosen by the IRS for an employment tax audit:

- You may be chosen for a general tax audit by the IRS computer.
- The IRS may receive complaints from disgruntled workers or even business competitors that you are misclassifying workers.
- You may be in an industry that has worker classification practices the IRS is targeting. In past years, the IRS has targeted hair salons, trucking firms, couriers, securities dealers, high technology firms, roofers, temporary employment agencies, nurses' registries, construction contractors, and manufacturers' representatives. If you're in a business or industry where classifying workers as ICs is common, the IRS is sure to target it sooner or later.
- You may have been chosen pursuant to the Employment Tax National Research Project, a program the IRS began in 2010, under which 6,000 taxpayers were selected over three years for comprehensive employment tax audits. The program's emphasis was worker misclassification and Form 1099 information reporting issues. The IRS plans to use the results to help determine guidelines for future employment tax audits, including proposing new penalties for misclassification.

- The IRS may be notified by a state agency that you have misclassified workers under state law. This most commonly occurs when terminated ICs apply for unemployment compensation and agency officials determine that the ICs should have been classified as employees under state law. Since 2007, the IRS has entered into agreements with 39 states to share the results of employment tax examinations. The IRS's Employment Tax Examination program (ETE) has examiners assigned to concentrate on employers suspected of worker misclassification based on information received from other federal and state government agencies.

- The IRS may decide to inspect your payroll tax records as part of its general enforcement program—similar to the random checks now conducted of passengers at airports. You'll be given seven days' notice, and an auditor will come to your office to conduct the inspection.

- You may come to the IRS's attention through an audit of an IC who hasn't been paying income or self-employment taxes.

- You may use an IC who receives only one 1099-MISC for a whole year. The IRS has a 1099 Matching Program targeting such individuals. The IRS also looks for individuals who receive both a Form W-2 and a 1099-MISC from the same employer in the same year.

- You may file more 1099 forms than average for companies in your industry, leading the IRS to conclude that you may be misclassifying your workers.

Audit Time Limits

As a general rule, the IRS has up to three years to audit a tax return after it's filed. Ordinarily, employment tax returns (Form 941) are filed every three months, but for purposes of the three-year limit, such returns are deemed filed on April 15 of the following year. However, employers who have employment tax liabilities of $1,000 or less for the year (average annual wages of $4,000 or less) file IRS Form 944, *Employer's Annual Federal Tax Return*, instead of Form 941. Form 944 is due on January 31, after the end of the tax year. The IRS has three years from that date to audit the employer.

If you have some workers you classify as employees and you file employment tax returns, the IRS will be able to conduct audits for only the previous three years. These are also known as open years. The IRS usually audits only one or two of these open years.

> CAUTION
> **Time limits are longer if you fail to file.** The audit time limit period—called a statute of limitations—starts to run only if you actually file an employment tax return. In other words, years in which you don't file an employment tax return are theoretically open to IRS scrutiny forever. If you've never hired employees, but have hired ICs, there is no time limit on a worker classification audit because you have never filed an employment tax return.
>
> In addition, there is no time limit on IRS audits if you filed a false or fraudulent return with the intent of evading taxes. This won't apply if a worker misclassification was due to an innocent or negligent mistake. But the IRS might invoke the fraud rule if it believes you knew the workers involved were employees and deliberately classified them as ICs to evade payroll taxes. In this event, the IRS may impose extremely harsh penalties. (See below.)
>
> Fortunately, the IRS has a general policy of not going back more than six years in conducting audits, even in cases of fraud or failure to file.

What the Auditor Does

As a first step, the IRS auditor will interview you. If your business is small—a sole proprietorship grossing less than $500,000 per year—the auditor will probably ask you to come to the local IRS office.

But if your business is a corporation, partnership, or sole proprietorship grossing more than $500,000 per year, the IRS will usually try to conduct the audit at your place of business. This is also known as a field audit. You're entitled to request that the audit be conducted elsewhere—for example, the IRS office or the office of your attorney or accountant. Explain that your business will be disrupted if the auditor comes there. It is usually a good idea to ward off on-site audits, because you don't want an IRS auditor snooping around your business premises.

The auditor will also try to talk to any workers he or she thinks might be misclassified. You can't prevent the auditor from doing this.

The auditor will seek detailed information about any workers you've classified as ICs, including their names and addresses, the services they performed, when they worked for you, how they were paid, whether you've filed 1099s reporting your payments, and whether you have signed independent contractor agreements.

IRS auditors have broad powers to inspect your records. To identify workers who may have been misclassified as ICs, IRS auditors often ask to see the following documents for the years being audited:

- copies of all IRS Form 1099s you've issued
- your payroll records and cash disbursement journals
- accounts payable records, and
- copies of all written contracts with outside workers to perform services on your behalf.

Among other things, IRS auditors will look for large payments to workers classified as ICs and try to trace whether payments have been made regularly over a number of years. Both may indicate employee status.

The IRS may also examine expense records for unreported or misreported payroll. The auditor may spot someone receiving money for services that have not been included on the employer's quarterly tax report, which will lead the auditor to look for payments to others in similar circumstances.

What You Need to Prove in an Audit

Unless the worker involved is a statutory IC (which is only possible for licensed real estate agents and direct sellers), the auditor will first attempt to determine whether you qualify for safe harbor protection. To qualify for the safe harbor, you must have:

- filed all required 1099 forms
- consistently treated all workers holding substantially similar positions as ICs for federal tax purposes, and
- had a good reason for treating the workers as ICs.

Most hiring firms are unable to meet these requirements. (See Chapter 3 for more information about statutory ICs and the safe harbor protection.)

If you don't qualify for safe harbor protection, the auditor will then determine whether the workers involved qualify as ICs or employees under the common law test. You'll need to convince the auditor that the workers are in business for themselves.

Examples of the information and documentation you should provide the auditor include:

- your signed IC agreements with the workers (see Chapter 12 for information on drafting independent contractor agreements)
- documentation provided by the workers showing that they're in business for themselves, such as proof of insurance, business cards and stationery, copies of advertisements, professional licenses, and copies of articles of incorporation
- a list of the workers' employees
- a list of the equipment and facilities owned by the workers
- the invoices the workers submitted for billing purposes
- the names and addresses of other companies the workers have worked for, preferably at the same time they worked for you, and
- copies of the 1099 forms you filed reporting the payments made to unincorporated workers to the IRS.

You should already have all this documentation and information in your IC files. (See Chapter 11 for information on IC files.) If you don't have it at hand, obtain it as quickly as possible.

If the workers do qualify as ICs under the common law test, the auditor's last task is to determine whether they are statutory employees—this status may apply only to corporate officers, home workers, life insurance salespeople, business-to-business salespeople, and drivers who distribute laundry, food, or beverages. (See Chapter 3 for more information about statutory employees.)

RESOURCE

Need more audit help? You can find detailed guidance on dealing with IRS auditors in *Tax Savvy for Small Business,* by Frederick W. Daily (Nolo).

Audit Results

After the audit ends, the IRS will mail you an examination report. This signals that the IRS considers the audit to be completed. Read the report carefully. If you're fortunate, you'll receive a "No Change Letter" stating that the auditor could not find sufficient basis for demanding any changes in your workers' status. On the other hand, if the auditor decides that you don't qualify for safe harbor protection and that you have misclassified workers under the common law test, the letter will demand that you change their status and it will impose assessments and penalties. However, if you qualify, the auditor may offer you a special settlement under the IRS Classification Settlement Program. In return for paying a reduced assessment, you must agree to treat the workers involved as employees in the future. (See below for more information on the settlement program.)

If assessments and penalties have been imposed, check to make sure they've been calculated correctly. (See below for more about IRS penalties.)

Appealing the Audit

Examination reports are not set in stone. You can fight back by negotiating with the auditor or, if that fails, by appealing to the IRS appeals office, the tax court, or the federal district court for your district.

Informal Negotiations

Before entering the world of appeals, you should informally negotiate with the IRS. Start with the person who audited you. It's not helpful to simply ask—or beg—the auditor to lower the amount of the assessments. Instead, you need to show the auditor that he or she was mistaken on the legal issues involved. You must show either that:

- you're entitled to safe harbor protection, or
- the auditor was mistaken in finding that the workers failed to qualify as ICs under the common law test.

If you can't get satisfaction from the auditor, you may have more success speaking with the auditor's manager. IRS managers are often more reasonable than field auditors.

Administrative Appeals

If informal negotiations don't work, you can appeal the examination report. One way to do this is to file an appeal with the IRS appeals office, a department of the IRS that is separate from the audit division. An IRS appeals officer will handle your administrative appeal. You may be given an opportunity to have a face-to-face meeting, or the entire matter may be handled by mail. Although the appeals office is a branch of the IRS, many taxpayers have been successful in appealing IRS audits there.

To begin the process of appealing, file a written protest with the local IRS district director, who will forward it to the appeals office. You must do this within 30 days of the date of the examination report or no later than 30 days of the date of an IRS 30-day letter. A 30-day letter will be sent to you if you fail to respond to the examination report within 30 days. It is your formal notice that your case is in dispute and you have 30 days to appeal. If you miss the deadline, you can still file an appeal in tax court or federal district court.

The IRS has a special program that allows companies to appeal employment tax issues on an expedited basis—before the IRS audit ends and you receive the auditor's final examination report. To do this, you must submit a written request to the case manager. There is no additional fee for an early appeal.

Tax Court Appeals

Whether or not you pursue an appeal to the IRS appeals office, you may appeal to the tax court. The tax court is a special federal court that just handles tax cases. Its procedures are less formal than those of the federal district courts. But what is most advantageous about tax court is that you may file an appeal without first paying the tax and penalties the IRS claims are due. This is not the case if you appeal in regular federal district court. (See below.)

To obtain tax court review of your case, you must file a petition with the court no later than 90 days after receiving an employment tax determination from the IRS. While your case is pending in tax court, you may start treating the workers whose employment status is involved—or those in similar positions—as employees rather than ICs. The tax court is not allowed to take this change into account in determining whether you properly classified the workers involved in your appeal as ICs.

Federal District Court Appeals

You also have the right to appeal in federal court. You can do this if you are dissatisfied with the outcome in the IRS administrative appeal, or you can forgo the administrative appeal and go straight to district court. But before you can appeal in court, you must pay the tax and penalties due and file a claim for refund with the IRS. After the IRS rejects the claim, you can sue for a refund in federal district court or federal claims court.

> SEE AN EXPERT
> **You have the right to be represented by a lawyer or CPA in an audit and during the administrative and court appeals process.** If you are facing the possibility of having to pay substantial assessments and penalties, it's probably sensible to hire professional help. An accountant, general business attorney, or others who own small businesses in your community may be able to refer you to a good local tax professional.

The Classification Settlement Program

When the IRS determines that hiring firms have misclassified employees as ICs, the Classification Settlement Program (CSP) gives the firms the chance to pay reduced assessments if they agree to classify the workers as employees in the future. The program is intended to encourage hiring firms to resolve worker classification cases as quickly as possible, saving themselves and the IRS time and money.

When the CSP Is Used

The CSP comes into play only if the IRS determines that you don't qualify for safe harbor protection and that the workers involved don't qualify as

ICs under the common law test. In this event, the examiner will determine whether you qualify for a CSP offer and make one if you do.

To qualify, you must have filed the required 1099 forms for the workers involved. However, your failure to file a small number of 1099 forms will not disqualify you from the CSP.

> **EXAMPLE:** The Acme Factory Outlet Store treated 150 workers as ICs. Acme filed all required 1099 forms for the workers except for three who were missed by the processing department. Acme's failure to file such a small number of 1099 forms does not disqualify it from the CSP.

CSP Offers

There are two types of CSP offers. Which one you receive depends on whether the examiner believes you may have satisfied the requirements for safe harbor protection.

CSP Settlement Offers		
1099s Timely Filed?	**Hiring Firm Entitled to Safe Harbor Protection?**	**Type of CSP Offer**
No	No	None
Yes	No	1 year's tax + employee treatment
Yes	Maybe	25% of 1 year's tax + employee treatment

One Full Year of Assessments

If the examiner concludes that you clearly don't qualify for safe harbor protection because you didn't consistently treat the workers involved as ICs or you clearly lacked a reasonable basis for the IC classification, your CSP offer will require you to pay the full IRS employment tax assessment for the workers involved for the year under audit. However, no IRS penalties will be imposed. You must also agree to begin treating the workers as employees.

EXAMPLE: The Acme Masonry Company hired two bricklayers in 2013. The two workers performed identical duties, but Acme treated one as an IC and the other as an employee—that is, it filed a 1099 for one and a W-2 for the other. The IRS audited the company. Acme is entitled to a CSP offer because it filed a 1099 for the worker it treated as an IC. However, it is absolutely clear that Acme is not entitled to safe harbor protection because it did not treat the similarly situated workers consistently. As a result, the CSP offer will be for one year's full employment tax assessments.

25% of One Year's Assessments

The CSP offer will require you to pay only 25% of one year's employment tax assessments if you've filed the required 1099 forms and if the IRS auditor concludes you might have a good argument that you're entitled to safe harbor protection. These will be cases where you have a problem establishing that you treated all your workers holding substantially similar positions as ICs, where it's not completely clear that you had a reasonable basis for classifying them that way, or both.

EXAMPLE: The IRS audits the owner of the Larkspur Dock and questions the status of several IC workers. The owner filed all required 1099 forms for the workers and consistently treated them all as ICs. The owner claims he had a reasonable basis for treating the workers as ICs because an accountant told him they qualified as such. The IRS auditor questions the accountant. It turns out the accountant gave the advice orally and can no longer remember what facts were provided. The owner has a potentially reasonable basis, but has not clearly established it because the advice was not in writing and the accountant can't remember why he gave it. Under these circumstances, the auditor proposes a CSP offer with a 25% assessment.

CSP Procedures

If you accept the CSP offer, the auditor will provide you with a standard closing agreement to sign. You will be required to begin treating the workers as employees on the first day of the calendar quarter following the date you sign the agreement. For example, if you sign the agreement on February 1, you'd have to begin employee treatment on April 1. Such treatment must extend not only to those workers who worked for you during the year under audit, but to all workers you hire subsequently to perform equivalent duties, regardless of their job titles.

You are free to reject a CSP offer if you wish. If you reject the offer initially, you can change your mind and accept it during the examination process.

Rejecting a CSP offer is not supposed to affect the outcome of the audit. However, the audit will be expanded to include all your open tax years instead of just one year.

To Accept or Not Accept a CSP Offer

Hiring firms that receive CSP offers often accept them. But this doesn't mean you should. If you conclude that you erred in classifying the workers involved as ICs, then the CSP offer may be a good deal. However, if you believe that you have a strong case for safe harbor protection or that the workers involved are ICs under the common law test, you may be better off rejecting the CSP offer and fighting the IRS. This may be particularly true if you provide generous pension or stock ownership benefits to your employees, or could become subject to Obamacare employer penalties in 2015 and later due to your failure to provide the workers involved with minimally adequate health insurance coverage (see Chapter 8). If you tell the IRS that the workers involved are really employees, the workers may sue you for employee benefits.

Accepting a CSP offer could cost you in another way as well. Once you accept the offer, you will have to treat the workers as employees for all future years. Even if the workers don't sue you for back benefits, you will feel the pinch in increased employment tax, workers' compensation expenses, unemployment insurance expenses, and health insurance costs. Indeed, these added expenses could cost you more than an appeal would have.

So, if you've got a leg to stand on, stand on it. Despite the cost of the appeal, you may end up saving money in the long run.

Voluntary Classification Settlement Program

In addition to the Classification Settlement Program, the IRS established the Voluntary Classification Settlement Program (VCSP) in 2011. The VCSP is similar to the CSP, but is designed for hiring firms that are not undergoing an IRS audit. As the name makes clear, the VCSP is completely voluntary.

In return for agreeing to reclassify independent contractors as employees in the future, the IRS grants the hiring firm full employment tax audit relief for all prior years and the firm need only make a minimal payment covering past employer payroll tax obligations.

Who Qualifies for the VCSP?

To be eligible for relief under the program, an applicant must:
- presently be treating the workers involved as independent contractors
- consistently have treated all workers in the same group or class as independent contractors in prior years (a class of workers includes all workers who perform the same or similar services)
- have filed all required Forms 1099-MISC for the workers for the previous three calendar years
- not currently be under an employment tax audit by the IRS, or a worker classification audit by the Department of Labor or a state agency, and
- have complied with any prior IRS or Department of Labor worker classification audit.

How to Apply

A hiring firm must apply for the VCSP and enter into a closing agreement with the IRS. To apply, a hiring firm must complete and file IRS Form 8952, Application for *Voluntary Classification Settlement Program*, at least 60 days before it wants to begin treating the workers involved as employees.

It's up to the hiring firm to select which class or group of workers it will reclassify as employees. For example, a construction firm that treats drywall installers, electricians, and plumbers as independent contractors can file Form 8952 to voluntarily reclassify the drywall installers as employees, while continuing to treat the electricians and plumbers as independent contractors.

What Happens If You're Accepted?

The IRS will review the completed Form 8952 to verify the applicant's eligibility. A hiring firm that is accepted will be required to sign a closing agreement with the IRS in which it agrees to:

- treat the workers involved and all future workers in the same class as employees for employment tax purposes, and
- pay the IRS an assessment equal to 10% of the amount of the total employment taxes that should have been paid for the workers the prior year. (This comes out to just over 1% of the wages paid to the reclassified workers for the previous year up to the applicable Social Security tax ceiling ($117,000 in 2014), and about .03% on wages over the ceiling.)

In return, the hiring firm won't be subject to an IRS employment tax worker classification audit of the workers involved for any prior years. As an additional sweetener, the IRS promises that it won't tell the U.S. Department of Labor or any state agencies about hiring firms that use the VCSP.

Is the VCSP a Good Deal?

On the face of it, the VCSP may look like a good deal. Particularly attractive is the unprecedentedly small amount the hiring firm must pay the IRS—little over 1% of payroll for the workers involved for the prior year, with no interest or penalties. This is substantially less than a hiring firm would have to pay if the IRS were to determine it misclassified the workers in an audit. In that event, the hiring firm would be liable for all the back FICA taxes it should have paid, a portion of the FICA tax it should have withheld, plus penalties and interest.

Is the VCSP a Good Deal? (continued)

However, before voluntarily agreeing to reclassify independent contractors as employees, a hiring firm should consider whether it qualifies for relief under the Section 530 safe harbor discussed above. Such relief is better than the VCSP because the hiring firm can continue to treat the workers involved as independent contractors for Social Security and Medicare tax purposes and need not pay any back employment taxes.

Any firm that qualifies for the VCSP will automatically satisfy two of the three requirements for relief under Section 530: filing 1099s and consistent treatment. The only additional requirement is showing some "reasonable basis" for treating the workers as employees at the time they were first classified as ICs.

If a hiring firm elects to use the VCSP and treat the workers involved as employees in the future, it will forever lose the right to raise the Section 530 safe harbor defense in an employment tax audit concerning such workers because it won't be able to satisfy the consistent treatment requirement. So this is not a step to be taken lightly.

Also, hiring firms should consider the nontax implications of agreeing to prospectively classify as employees workers who have been treated as independent contractors (perhaps for years). For example, the affected workers may claim that they are entitled to employee benefits for previous years, such as pension and health benefits. Larger employers may have to provide such employees with health benefits or face stiff penalties under Obamacare during 2015 and later (see Chapter 8).

The VCSP might be a good deal for a hiring firm that has flagrantly misclassified as independent contractors workers who are clearly employees under the IRS common law test.

But it's likely that only a small minority of hiring firms are in the flagrant-misclassification position. In most cases where worker classification questions arise, it's unclear how the workers should be classified under the infamous IRS 20-factor test, which centers on whether the hiring firm has the right to control the worker. The Section 530 safe harbor is designed to help hiring firms in these situations.

For these reasons, relatively few hiring firms have taken advantage of the IRS's VCSP.

IRS Assessments for Worker Misclassification

The assessments the IRS can impose for worker misclassification vary enormously, depending upon whether the IRS views your misclassification as intentional or unintentional. In some cases, assessments for misclassification could be 40% of gross payroll or more. The most strict penalties, of course, are imposed for intentional misclassification—where you knew the workers were your employees but classified them as ICs anyway to avoid payroll taxes. The IRS will likely conclude your misclassification was intentional if you admit you knew the workers were employees, or if it would have been clear to any reasonable person that the workers were employees under the common law test.

> **EXAMPLE:** Bolo Press, a publisher of sports books, has a six-person production department. Bolo reclassifies its production employees as ICs so it can stop paying payroll taxes for them. After the reclassification, Bolo's owners treat the production workers just the same as when they were classified as employees—they tell the workers what time to come in and leave, closely supervise their work, and give them fringe benefits, such as health insurance and pension benefits. The production workers work solely for Bolo and make no attempt to market their services elsewhere.
>
> In an audit, the IRS concludes that the workers are employees. IRS auditors would likely conclude that the misclassification was intentional. Bolo clearly knew the production workers were really employees and reclassified them as ICs in disregard of the law simply to avoid paying payroll taxes.

On the other hand, worker misclassification is unintentional if you believed in good faith, though mistakenly, that the workers were ICs. This can involve tricky semantics. The common law test for worker classification used by the IRS is complex and difficult to apply and often doesn't provide a conclusive answer about how to classify a worker.

Many workers fall into a gray area where it is unclear how to classify them. If you can show that some of the common law factors indicate IC status, your misclassification should be regarded as unintentional. You should be able to do this in all but the most blatant misclassification cases.

Penalties for Unintentional Misclassification

When you hire an unincorporated IC, you are generally required to report the payments made to the worker on IRS Form 1099-MISC. There are two ranges of assessments the IRS may impose for unintentional worker misclassification: One is imposed if you filed all required 1099 forms for the workers the IRS claims you misclassified, and the other is imposed if you did not file the 1099 forms.

1099 Forms Filed

If you filed 1099 forms, you will be required to pay each of the following.

FICA taxes that should have been withheld. You'll have to pay 20% of the FICA taxes (Social Security and Medicare) the employees should have had withheld from their pay—that is, 1.24% of the misclassified workers' wages up to the FICA Social Security tax ceiling. (For 2011 and 2012 only, the employee's Social Security tax was reduced by two percentage points to 4.2%—thus, the 20% penalty comes to 0.84% of the workers' wages rather than 1.24%.) In addition, you will have to pay 0.29% of all the workers' wages to make up for lack of Medicare tax withholding.

FICA taxes you should have paid. You'll have to pay 100% of the FICA taxes you should have paid on the workers' behalf as their employer—that is, 6.2% of the employees' wages up to the Social Security tax ceiling, plus 1.45% of all the employees' wages.

Penalty for failure to withhold income tax. You will owe 1.5% of all the wages that were paid to the misclassified workers—a penalty for your failure to withhold federal income taxes from their paychecks.

FUTA taxes that should have been paid. You will lastly owe all FUTA taxes (federal unemployment taxes) that should have been paid. The FUTA tax rate is 6% of the first $7,000 in employee wages, or 0.6% of the first $7,000 if the applicable state unemployment tax was timely paid. (If a worker was paid $7,000 or more for a year, this amounts to either $420 or $42.)

FICA Social Security Tax Ceiling

FICA taxes consist of two separate taxes: a 6.2% Social Security tax and a 1.45% Medicare tax on both the employer and employee. For 2011 and 2012 only, the employee's share of the Social Security tax was reduced by two percentage points to 4.2%. There is a ceiling on the Social Security tax—that is, a salary level beyond which the tax need not be paid. The ceiling increases every year. In 2014, the ceiling was $117,000. However, the Medicare tax must be paid on all compensation paid to an employee.

You must pay the full assessments for your failure to withhold employee FICA and income taxes from misclassified workers' compensation even if the workers paid all these taxes themselves. This means that the IRS could end up collecting more tax than would have been due had you classified the workers as employees and paid payroll taxes.

Together, these assessments amount to 16.86% of the compensation you paid each misclassified worker up to the $7,000 FUTA tax ceiling and then 10.68% of compensation up to the FICA Social Security tax ceiling. Any payments to a worker in excess of the Social Security tax ceiling are assessed at a 3.24% rate.

EXAMPLE: The IRS decides that Acme Sandblasting Corporation unintentionally misclassified five workers as ICs during 2013. Acme paid each worker $20,000 during that year and reported the payments on Form 1099. Based on this $100,000 in payments, the IRS assessment would be $12,780. This is calculated as follows:

20% of employee FICA tax	= $1,530
100% of employer FICA tax	= $7,650
1.5% of all employee wages	= $1,500
6% FUTA tax for five workers, each paid $20,000	= $2,100

1099 Forms Not Filed

If you failed to file the 1099 forms, the employee FICA and income tax assessments double. You must pay each of the following.

FICA taxes that should have been withheld. You'll have to pay 40% of the FICA (Social Security and Medicare) taxes the employees should have had withheld—that is, 2.48% of the misclassified workers' wages up to the FICA Social Security tax ceiling. (For 2011 and 2012 only, the employee's Social Security tax was reduced by two percentage points to 4.2%—thus, the 40% penalty comes to 1.68% of the workers' wages rather than 2.48%.) In addition, you will have to pay 0.58% of all the workers' wages to make up for the lack of Medicare tax withholding.

FICA taxes you should have paid. You'll have to pay 100% of the FICA taxes you should have paid on the misclassified workers' behalf as their employer—that is, 6.2% of the employees' wages up to the Social Security tax ceiling, plus 1.45% of all the employees' wages.

Penalty for failure to withhold. You will owe 3% of all the wages that were paid to each misclassified worker as a penalty for your failure to withhold federal income taxes from the workers' paychecks.

FUTA taxes that should have been paid. You will lastly owe all FUTA taxes that should have been paid—6% or 0.6% of the first $7,000 in compensation, depending on whether you paid state unemployment taxes.

> **EXAMPLE:** Recall the example above, in which $12,780 was assessed on $100,000 in payments to five misclassified workers. If you did not file Form 1099 for those workers, you'd have to pay $15,810. This would break down as follows:
>
> | | | |
> |---|---|---|
> | 40% of employee FICA tax | = | $3,060 |
> | 100% of employer FICA tax | = | $7,650 |
> | 3% of all employee wages | = | $3,000 |
> | 6% FUTA tax for five workers, each paid $20,000 | = | $2,100 |

> ### Pre-2011 FUTA Tax Rate
>
> For many years, the FUTA tax rate was 6.2% of the first $7,000 of employee wages. However, starting July 1, 2011, the rate was decreased to 6% of the first $7,000 of wages. IRS audits dealing with tax years 2011 or earlier must take into account the higher rate in determining the FUTA assessments due.

Penalties for Intentional Misclassification

IRS assessments are far higher if the IRS concludes that you intentionally misclassified workers as ICs. You will be required to pay out of your own pocket all the FICA tax that you should have withheld from the employees' paychecks. You must pay:

- 100% of the FICA (Social Security and Medicare) taxes the misclassified workers should have had withheld—that is, 7.65% of the employees' wages subject to FICA (for 2011 and 2012 only, the penalty is 5.65% due to the 2% reduction in the employee's share of the Social Security tax), plus
- 100% of the FICA taxes you should have paid on the workers' behalf as their employer—that is, 7.65% of the employees' wages subject to FICA, plus
- 20% of all the wages that were paid to the workers to make up for your failure to withhold federal income taxes from their paychecks, plus
- all FUTA taxes that should have been paid—6% of the first $7,000 in worker compensation or 0.8% if state unemployment taxes were paid.

Together, these assessments amount to a whopping 41.5% of worker compensation up to the $7,000 FUTA tax ceiling and then 35.3% of payments up to the Social Security tax ceiling.

Comparing these assessments with those that can be imposed for unintentional misclassification is a sobering exercise. As illustrated above, if you paid $100,000 to five workers the IRS claimed you unintentionally misclassified as employees during 2013, the assessments would total $12,780 if you filed 1099 forms, or $15,810 if you failed to file 1099 forms. But if the IRS claims you intentionally misclassified the workers, the assessments would total $37,400. This breaks down as follows:

100% of employee FICA tax	=	$7,650
100% of employer FICA tax	=	$7,650
20% of all employee wages	=	$20,000
6% FUTA tax for five workers who were each paid $20,000	=	$2,100

Offsets for Worker Income Tax Payments

The only good thing about the intentional misclassification assessment is that the income tax portion can be reduced if you can prove that the misclassified worker paid his or her income taxes for the years in question. Such a reduction is called an offset or abatement.

The IRS will not help you prove income taxes were paid. IRS examiners will not request that workers provide copies of their income tax returns nor will the IRS give these returns to you. Instead, you need to file IRS Form 4669, *Statement of Payments Received*. This form states how much tax the worker paid on the wages. The worker must sign the form under penalty of perjury. You must file a Form 4669 for each worker involved along with Form 4670, *Request for Relief From Payment of Income Tax Withholding*, which is used to summarize and transmit the Form 4669.

The IRS examiner has the discretion to accept Forms 4669 and 4670 before the examination is closed and reduce the assessment. Otherwise, you must file them with your IRS service center.

> EXAMPLE: The IRS determines that Acme Sandblasting Corporation intentionally misclassified a computer consultant as an IC in 2013. Acme paid the worker $100,000. The IRS assessment is $30,848.40. However, the consultant paid all income taxes due on her compensation. Acme has the consultant sign IRS Form 4669 and submits it to the examiner along with Form 4670 before the IRS examination is

closed. The examiner wipes out the entire 20% income tax penalty. The assessment is reduced by $20,000.

Offsets for Worker FICA Tax Payments

Theoretically, if the misclassified workers paid their FICA taxes, you may also be entitled to an offset of the employee FICA portion of the assessment. However, in practice, this offset is difficult or impossible to obtain because a misclassified worker has a right to claim a refund for all the self-employment taxes he or she paid for the years covered by the audit. If the worker claims the refund, you can't get the offset. Because of this right, the IRS will not give you an employee FICA offset unless the misclassified worker fails to file a claim for a refund of these taxes within the statutory time limit, either two or three years.

Intentional Misclassification Assessments May Be Lower

Oddly, if you are able to get an offset for the income tax portion of an intentional misclassification assessment, the final assessment may be less than what you would have had to pay for an unintentional misclassification. However, you are not allowed to choose which assessment rules will be used. If your misclassification was unintentional, the assessment rules will apply and you will not be entitled to any offsets. Some hiring firms have actually attempted to convince the IRS that a misclassification was intentional so that they could obtain assessment offsets.

Penalties for Worker Misclassification

In addition to the assessments discussed above, the IRS has the option of imposing an array of other penalties on companies that misclassify workers.

Trust Fund Recovery Penalty

As far as the IRS is concerned, an employer's most important duty is to withhold FICA and income taxes from its employees' paychecks and pay the money to the IRS. Employee FICA and federal income taxes are also known as trust fund taxes because the employer is deemed to hold the withheld funds in trust for the U.S. government. The IRS considers failure to pay trust fund taxes to be a very serious transgression.

The IRS may impose a penalty known as the trust fund recovery penalty, formerly called the 100% penalty, against individual employers or other people associated with the business. (Internal Revenue Code (IRC) § 6672.) These are people the IRS deems responsible for failing to withhold employee FICA and federal income taxes and pay the withheld sums to the IRS. Failure to pay payroll taxes is willful if you knew the taxes were due and didn't pay them. The IRS will conclude you have acted willfully if you should have known the workers involved were employees, not ICs.

The trust fund recovery penalty is also known as the 100% penalty because the amount of the penalty is equal to 100% of the total amount of employee FICA and federal income taxes the employer failed to withhold and pay to the IRS. This can be a staggering sum.

> **EXAMPLE:** The IRS determines that Acme Sandblasting Corporation intentionally misclassified a computer consultant as an IC in 2010. Acme paid the consultant $100,000. The IRS decides to impose the trust fund recovery penalty against Acme. Acme should have withheld and paid to the IRS $7,650 in employee FICA taxes and withheld $25,000 in federal income taxes, for a total of $32,650 in trust fund taxes. The 100% penalty is $32,650.

Liability for the 100% Penalty

If you're a business owner, you'll be personally liable for the 100% penalty—in other words, you will have to pay it out of your own pocket. Business owners include sole proprietors, general partners, and corporate officers such as the president, vice president, secretary, and treasurer, whether or not they own any stock.

However, the scariest thing about the trust fund recovery penalty is that nonowner employees, such as office managers, accountants, bookkeepers, and even some clerks, may also be held personally liable for it. They may be on the hook if the IRS concludes they willfully prevented the IRS from collecting the unpaid payroll taxes—in other words, they knew the taxes were due and didn't do anything about it. Most vulnerable are those who:

- made the business's financial decisions
- had authority to sign checks
- had the power to decide which bills to pay, or
- signed the business's payroll tax returns, such as the quarterly IRS Form 941.

Appealing a Penalty

If an IRS revenue officer decides you are responsible, you will be sent a notice and tax bill. You are entitled to appeal the penalty and have a hearing before an IRS appeals officer. You must file an appeal within 30 days.

Other Penalties

The IRS has the option of imposing many other penalties on hiring firms that misclassify workers. These include:

- A $100 penalty for each W-2 you failed to send a misclassified employee. If the failure to file was intentional, the penalty is the greater of $100 or 10% of the amount not reported correctly on each W-2 form. (IRC § 6722.)
- A $100 penalty for each 1099 form you failed to file. (IRC § 6721.)
- If the IRS determines you intentionally disregarded the rules requiring 1099 forms to be filed, a penalty equal to the greater of $250 or 10% of the compensation paid the worker. (IRC § 6721(e).)
- If employment tax returns were not filed, a delinquency penalty of 5% per month up to 25% of the tax determined to be due. (IRC § 6651.)
- For failing to deposit the taxes found to be due, a penalty of up to 15% of the undeposited taxes. (IRC § 6656.)

- For negligently or intentionally disregarding IRS rules and regulations, a penalty of up to 20% of the underpayment that is due to
  the negligence. (IRC § 6662.)
- A fraud penalty of 75% of the underpayment if the IRS determines
  that the underpayment is due to fraud; no negligence penalty is
  imposed in this event. (IRC § 6663.) Fraud is intentional wrong-
  doing designed to evade a tax you believed you owed.

Generally, the more severe penalties are imposed only where you
intentionally misclassified workers.

Interest Assessments

The IRS can impose interest on employment tax assessments and penalties.
The interest rate is adjusted every three months and compounded daily. It
is currently about 3%.

The Bottom Line: All You Could Owe

Factoring in all these assessments, penalties, and interest, you can make a rough
estimate of what you'll have to pay.

If the IRS determines you unintentionally misclassified a worker for whom
you filed all required 1099 forms, you'll have to pay about 20¢ for every dollar
you paid the worker, and 25¢ for every dollar if you didn't file 1099 forms. But
if the IRS finds your misclassification intentional, you'll have to pay about 50¢
for each dollar you paid the worker.

Criminal Sanctions

In rare cases, the IRS may conduct a criminal investigation and have the
U.S. Justice Department prosecute. Criminal fines and even jail time can
be imposed if you're convicted of tax fraud.

Obamacare Penalties

Starting in 2015, "large employers" (those with 100 or more full-time or full-time-equivalent employees in 2015, and 50 or more in 2016 and later) will become subject to severe penalties if they fail to provide their full-time employees with minimally adequate affordable health insurance. Large employers who are found to have misclassified workers as ICs to avoid Obamacare's requirements could be subject to penalties of $2,000 per year for each full-time employee without coverage (subject to certain reductions). See Chapter 8 for a detailed discussion of Obamacare.

Retirement Plan Audits

If your company has a retirement plan, you should be concerned about IRS retirement plan audits. Retirement plans are not audited as part of an ordinary business or employment tax audit. Instead, the IRS has specially trained revenue agents in every district just for retirement plan audits. This type of audit may derive from a prior business audit or from a review of annual IRS tax reporting Form 5500, which is required for most retirement plans.

Companies with tax-qualified retirement plans may have special problems with retirement plan audits if they classify workers as ICs.

Tax-Qualified Retirement Plans

A tax-qualified retirement plan is a retirement plan that covers business owners and employees and satisfies the requirements of the federal Employee Retirement Income Security Act (ERISA). That law is enforced by the U.S. Department of Labor, the IRS, and the Pension Benefit Guarantee Corporation.

Contributions to a tax-qualified retirement plan are tax deductible by the business, as are contributions by participating employees. In addition, income from retirement plan investments is tax free until it is withdrawn by the plan participants.

Antidiscrimination Rules

You must satisfy complex ERISA rules to obtain these tax benefits. The most important are antidiscrimination requirements providing that the principal owners of a business cannot provide benefits only to themselves, corporate officers, or highly paid employees. If there are other employees, many, but not all, must be included in the plan as well. In general, employees don't have to be covered from the moment they're hired—but if they're employed long enough and are old enough, you have to bring them into the plan.

ERISA has even more complex rules concerning which workers must be counted under the antidiscrimination rules and how many need to be covered. The antidiscrimination rules are satisfied, for example, if 70% of employees are covered. There are other ways to satisfy the rules that may require fewer employees to be included in the plan.

Losing Tax-Qualified Status

If the IRS determines that workers you classify as ICs are really employees for ERISA purposes, it's possible that not enough employees will be covered to satisfy the antidiscrimination rules. This means that your pension plan could lose its tax-qualified status. In this event, all previous tax deductions for benefits or contributions to the plan can be thrown out. Your business can lose the deductions, and the benefit recipients will have to pay taxes on the benefits.

> **EXAMPLE:** Acme Sandblasting Corporation has a tax-qualified retirement plan. Acme has 100 employees and another 100 workers classified as ICs. Seventy of the 100 employees are covered by the pension plan, apparently satisfying the antidiscrimination rules because 70% of workers Acme classifies as employees are covered. However, the IRS determines that the 100 ICs should be classified as employees for ERISA purposes. Acme really has 200 employees and 140 had to be covered. As a result, Acme's retirement plan loses its tax-qualified status.

The IRS uses the common law right-of-control test to determine whether workers are employees or ICs for ERISA purposes. If a worker qualifies as an employee for federal payroll tax purposes, he or she is an employee for ERISA purposes as well.

> **SEE AN EXPERT**
>
> **Get help for retirement plan issues.** This is an extraordinarily complex area of the law, beyond the competence of most attorneys, let alone laypeople. You should discuss this issue with your retirement plan administrator or seek advice from a retirement plan consultant or an attorney or CPA specializing in this field.

Worker Lawsuits for Pensions and Other Benefits

A company that provides its employees with pensions, stock options, and other benefits faces the possibility that workers it has incorrectly classified as ICs will file expensive lawsuits asserting that they are really employees and therefore entitled to the benefits. This happened in a highly publicized case involving the Microsoft Corporation in which the company was sued by several workers it had improperly classified as ICs for federal payroll tax purposes. After lengthy litigation, the court held that the workers were entitled to full employee benefits for the entire time they had worked for Microsoft, including coverage under Microsoft's discount stock purchase plan and 401(k) plan. (*Vizcaino v. Microsoft,* 173 F.3d 713 (9th Cir. 1999).)

Companies that offer generous employee retirement or stock options plans like Microsoft's can avoid the problems Microsoft encountered by making sure that their plan eligibility provisions explicitly exclude workers whom the company has classified as ICs or as contract employees. Coverage should not depend on how the IRS or any other agency classifies the workers. The workers should sign IC agreements in which they waive any claims to such benefits. Companies should talk with their benefit plan administrators to make sure their plans contain such language. If the plans don't have such language, then the plans should be amended to add it.

State Payroll Taxes

State Unemployment Compensation 130

    Unemployment Compensation Audits 131

    Threshold Requirements for UC Taxes 133

State UC Classification Tests 134

    The Common Law Test 136

    The ABC Test 137

    Modified ABC Tests 141

    Other Tests 142

    Domestic Service 144

    Agricultural Labor 144

    Special Rules for the Construction Industry 145

    Statutory Employees 147

    Exemptions From Coverage 148

State Disability Insurance 150

State Income Taxes 153

E mployers in all states must pay and withhold state payroll taxes for employees. These taxes include:

- state unemployment taxes in all states
- state income tax withholding in most states, and
- state disability taxes in a few states.

You do not have to withhold state payroll taxes for ICs. Thus, whenever you hire a worker, you must decide whether he or she is an employee or IC under your state's payroll tax laws. This isn't a decision to be taken lightly, because businesses are more likely to face a state payroll tax audit than any other type.

Many states have stepped up their worker classification enforcement efforts in recent years. For example, New York State established a Joint Enforcement Strike Force to address the issue of employee misclassification. The task force facilitates the sharing of information among various state agencies. Employers who misclassify workers are investigated and may be subject to failure-to-file penalties, fraud penalties, higher tax assessments and tax rates, and periodic audits. In some cases, state investigators are conducting surprise raids on companies after getting tips from employees.

RESOURCE

Don't forget workers' compensation. In addition to state payroll taxes, you must purchase workers' compensation insurance for employees (but not for independent contractors). Often, this involves analyzing workers under yet another test. See Chapter 6 for more information.

State Unemployment Compensation

Federal law requires all states to create an unemployment insurance fund. Employers are required to pay into this fund for their employees. Employees make no contributions, except in Alaska, New Jersey, Pennsylvania, and Rhode Island, where employers withhold small contributions from their employees' paychecks.

Unemployment compensation (UC) is only for employees—ICs cannot collect it. Firms that hire ICs don't have to pay unemployment compensation taxes for them. This is one of the significant benefits of classifying workers as ICs: Businesses can save the hundreds of dollars per year they would have to pay for each worker if he or she were an employee.

> ## The Cost of Unemployment Compensation Insurance
>
> The unemployment tax rate varies from state to state and depends partly on the age of the business, the industry, and how many claims have been filed by the company's employees. Employers who maintain a stable payroll and file and pay their unemployment taxes on time will generally have a lower unemployment tax rate than those with high turnovers or large fluctuations in their payroll and those who don't file or pay their taxes on time.
>
> As a general rule, however, the unemployment tax rate is usually somewhere between 2% to 5% of wages, up to the maximum amount of wages that are taxable under the state's unemployment compensation law.

Unemployment Compensation Audits

You're more likely to be audited by a state UC auditor than by any other type of government auditor, including those with the IRS. There are two main reasons for this. First, most states have become very aggressive in auditing hiring firms for UC purposes. The more workers that are classified as ICs for UC purposes, the less money there is for the state's UC fund—and states are increasingly more aggressive about guarding these funds. In addition, state unemployment auditors are often the first to become aware of a firm's worker classification practices because ICs often apply for unemployment compensation when their work for a business ends. (There is nothing to prevent workers you have treated as ICs from walking into the local unemployment insurance office and applying for employee unemployment benefits after their work for you or another hiring firm has ended.) UC audits can also be triggered by referrals from other agencies, comparison of payroll and workers' compensation records, and tips that an employer may be in violation of law (often from competing businesses or disgruntled workers).

When a worker classified as an IC files an unemployment compensation application, state unemployment auditors will investigate the hiring firm to determine whether the worker should have been treated as an employee under the state's unemployment compensation law. If the state auditors determine the worker should have been classified as an employee, they will require the company to pay all the unemployment taxes it should have paid for the worker going back several years (three years is common), plus interest.

In addition, auditors usually impose penalties for the misclassification. Penalties vary from state to state. A 10% penalty is common, but penalties are much higher in some states. For example, if a California employer willfully misclassifies a worker as an IC, the state can impose a penalty equal to about 50% of the total compensation paid the worker for the prior three years.

To add insult to injury, if other workers do similar work for the same company, the employer will have to pay UC taxes for those employees as well. The misclassified employees will also be eligible for workers' compensation benefits (see Chapter 6) and any fringe benefits the company provides its employees, such as overtime, health insurance, and retirement plans.

Unfortunately for businesses, unemployment compensation agencies in a great many states share information with other state agencies and the IRS. For example, they inform other agencies that a worker was misclassified for UC purposes. The IRS and other agencies will likely assume that the worker has been misclassified for their purposes as well and conduct an audit. An unemployment compensation audit may only be the first of many audits: Workers' compensation, state income tax, and IRS audits may well follow. Clearly, deciding how to classify a worker for unemployment compensation purposes is a very important decision for any business.

Threshold Requirements for UC Taxes

Before going to the time and trouble of trying to decide whether workers are employees or ICs under your state UC law, first see whether you're required to pay for UC coverage for your employees. In most states, if your payroll is very small, you won't have to pay UC taxes regardless of how your workers are classified.

In about half of the states, you must pay state UC taxes for employees if you're paying federal UC taxes, also called FUTA taxes. This means you must pay state UC taxes if:

- you pay $1,500 or more to employees during any calendar quarter—that is, any three-month period, or
- you have at least one employee during any day of the week for 20 weeks in a calendar year (the 20 weeks need not be consecutive).

Other states have payroll or service requirements that are stricter than FUTA requirements. For example, California employers must pay UC taxes if they pay one or more employees $100 or more per quarter, while Michigan employers must pay UC taxes for any employee who is paid $1,000 or more in a calendar year. And several states provide the broadest possible UC coverage by requiring employers to pay UC taxes for all employees. The requirements of the states with stricter-than-FUTA requirements are summarized in the following chart. In these states, an employer must pay UC taxes for an employee who works for the period of time or receives the amount of compensation reflected in the right-hand column.

State UC Tax Requirements	
State	**Minimum Period of Time or Payroll**
Alaska	Any time
Arkansas	One employee for ten or more days in a calendar year
California	Over $100 in a quarter
Hawaii	Any time
Maryland	Any time
Massachusetts	13 weeks or $1,500 in a quarter
Michigan	20 weeks or $1,000 in a calendar year
Minnesota	Any time
Montana	$1,000 in the current or preceding year
Nevada	$225 in a quarter
New Jersey	$1,000 in a year
New Mexico	20 weeks or $450 in a quarter
New York	$300 in a quarter
Oregon	18 weeks or $1,000 in a quarter
Pennsylvania	Any time
Rhode Island	Any time
Utah	Any time
Washington	Any time
Wyoming	Any time

State UC Classification Tests

Each state has its own unemployment compensation law administered by a state agency, often called the department of labor. Each state's law defines who is and who is not an employee for unemployment compensation purposes. Definitions for almost all states fall into one of three categories:

- the common law test
- a three-part ABC test, or
- a modified ABC test.

Find the category for your state on the list below and then read the appropriate discussion, also below. This should give you a general idea of whether a particular worker is an employee or IC for UC purposes. For more detailed information, contact the unemployment compensation agency in your state. Most of these agencies have websites and free information pamphlets. Montana, Wisconsin, and Wyoming do not fit into any of the three standard categories, and are covered separately below.

Which State's UC Law Applies?

Businesses sometimes have to determine which state's UC law applies to a worker. For example, if a hiring firm based in California hires a worker in Wisconsin to perform services, does the California or Wisconsin UC law apply?

In general, workers are covered by the UC law of the state in which the work is performed. However, depending on the circumstances, as many as four tests are used to determine the applicable law. The tests are applied in the following order until a determination is made.

Test 1: Localization

First, the UC law of the state in which a worker's services are localized applies. A worker's services are "localized" in the state in which all or most of the worker's services are performed. A worker may perform incidental services elsewhere, as where the out-of-state service is temporary or transient or consists of isolated transactions. In that case, the law of the state where the worker performs most of his or her services controls.

Test 2: Base of Operations

If Test 1 doesn't apply because there isn't a single state of "localization" for the worker, then the UC law of the state in which the worker has his or her base of operations applies if he or she performs some service in that state. "Base of operations" is a more or less permanent place from which the worker starts work and customarily returns to receive instructions or communications from customers or others, replenishes stocks or supplies, repairs equipment, or performs other functions relating to the rendition of services.

Which State's UC Law Applies? (continued)

Test 3: Place of Direction and Control

If Tests 1 and 2 don't apply, the UC law of the state from which the worker's services are directed and controlled governs.

Test 4: Worker's Residence

If Tests 1 through 3 don't apply, the UC law of the state in which the worker resides applies if he or she performs some services there.

The Common Law Test

Many of the most populous states—including California, Florida, New York, and Texas—use some version of the common law test to determine whether a worker is an employee for UC purposes. Under this test, a worker is an employee if the person for whom he or she works has the right to direct and control how the work is performed, including when, where, and how the work is to be done.

This is the same test that the IRS uses to determine whether a worker is an employee or IC for federal unemployment tax (FUTA) purposes, although not all states use the test in exactly the same way. (See Chapter 3 for a discussion of the common law test as applied by the IRS.)

Generally, any worker who qualifies as an IC under the IRS's version of the common law test will also be an IC under a state unemployment agency's common law test. However, there are no guarantees. Classifying a worker under the common law test is far from an exact science and, particularly in borderline cases, opinions may differ. State unemployment agencies are not bound by an IRS determination that a worker is an IC, so it is possible that that worker could be deemed an IC by the IRS and an employee by the hiring firm's state unemployment compensation agency. It's a good idea to obtain detailed information about how your state unemployment agency applies the common law test.

The ABC Test

Over half the states use a statutory test written by their legislatures to determine whether a worker is an employee or an IC for unemployment compensation purposes. This is called the ABC test because it contains three parts.

To prove that a worker is an IC for state unemployment compensation purposes, you must satisfy all three prongs of the ABC test. You must show that the worker:

- is free from your control or direction in performing the services, both in any oral or written contract of service and in reality
- provides services that are either outside your company's usual course of business, or performed outside of your place of business, and
- is carrying on an independently established trade, occupation, profession, or business

If any one of the three prongs of the ABC test is not satisfied, the worker will be classified as an employee for unemployment compensation purposes—and you must pay state unemployment compensation taxes for that individual.

The Strictest Test Around

The full-blown ABC test is the most strict worker classification test. It's possible for a worker to qualify as an IC for IRS and other purposes under the less strict common law test and be an employee under the ABC test for state unemployment compensation purposes. In this situation, you wouldn't withhold or pay federal payroll taxes, but would pay state UC taxes.

But this poses practical problems. Paying UC taxes for a worker makes the worker look like your employee. You could attempt to explain to an IRS auditor that your state has an extremely strict ABC test, but the auditor will still likely view payment of state UC taxes for a worker as a strong indicator of employee status. The safest course is to classify a worker who fails the ABC test as an employee for all purposes.

Prong A: Control or Direction of the Work

The first part of the ABC test, Prong A, requires that you not have the right to exercise control or direction over the worker's services. Your control must be limited to accepting or rejecting the results the worker achieves, not how he or she achieves them. This is simply a restatement of the common law right-of-control test—the test used to determine worker status for IRS and many other purposes. (See Chapter 3 for a discussion of the IRS test.)

The factors state UC auditors examine to determine whether you have the right to control a worker differ somewhat from state to state and are determined by state UC laws, regulations, and court rulings. For example, Maryland regulations provide that a worker is considered to be free from a hiring firm's direction or control if the firm does not:

- require the worker to comply with detailed instructions about when, where, and how the person is to work
- train the worker
- establish set hours of work
- establish a schedule or routine for the worker, or
- have the power to fire the worker for failing to obey specific instructions. (Md. Regs. Code title 24, § 24.02.01.18(B)(3)(a).)

To find out the details of your state's test, you will have to contact the agency in your state that enforces unemployment compensation laws.

Prong B: Outside Service

The second part of the ABC test focuses on whether the worker's services are outside of your normal business. Prong B is satisfied if either:

- the worker's service is outside the usual course of your business operations, or
- the work is performed completely outside your usual places of business.

Usual course of business operations. State auditors seek to determine whether the worker's services are an integral part of—that is, closely related to—your normal daily business operations. You're likely to exercise control over such workers because they are so important to your business's success or continuation.

State Tests for Unemployment Compensation			
State	**Test**	**State**	**Test**
Alabama	Common law	**Montana**	Other
Alaska	ABC	**Nebraska**	ABC
Arizona	Common law	**Nevada**	ABC
Arkansas	ABC	**New Hampshire**	ABC
California	Common law or ABC	**New Jersey**	ABC
Colorado	Modified ABC (AC)	**New Mexico**	ABC
Connecticut	ABC	**New York**	Common law
Delaware	ABC	**North Carolina**	Common law
Dist. of Columbia	Common law	**North Dakota**	Common law
Florida	Common law or ABC	**Ohio**	Common law
Georgia	Modified ABC (AC)	**Oklahoma**	ABC
Hawaii	ABC	**Oregon**	Modified ABC (AC)
Idaho	Modified ABC (AC)	**Pennsylvania**	Modified ABC (AC)
Illinois	ABC	**Rhode Island**	ABC
Indiana	ABC or common law	**South Carolina**	Common law
Iowa	Common law	**South Dakota**	Modified ABC (AC)
Kansas	Modified ABC (A)	**Tennessee**	ABC
Kentucky	Common law	**Texas**	Common law
Louisiana	ABC	**Utah**	Modified ABC (AC)
Maine	ABC	**Vermont**	ABC
Maryland	ABC	**Virginia**	Common law
Massachusetts	ABC	**Washington**	ABC plus three
Michigan	Common law	**West Virginia**	ABC
Minnesota	Common law	**Wisconsin**	Other
Mississippi	Common law	**Wyoming**	Other
Missouri	Common law		

EXAMPLE 1: Jeremy works part time on the assembly line at the General Widget Corp. Jeremy helps assemble widgets, which is what General Widget does. Jeremy's services are essential to the nature of General Widget's business because it can't produce widgets without people working on the assembly line. Prong B isn't met because Jeremy does not work outside the hiring firm's normal course of business.

EXAMPLE 2: General Widget Corp. hires Jessica, an attorney with her own practice, to defend it in a products liability lawsuit. General Widget is in the business of producing widgets, not defending lawsuits, so Jessica's services are outside General's normal business. Prong B is satisfied.

Work performed outside the business. Even if a worker's services are an integral part of your business operations, you can still satisfy Prong B if the services are performed outside your place of business. In other words, the worker must not work on any of your business premises.

EXAMPLE: General Widget Corp. contracts with Arnie to provide important component parts for its widgets. Arnie builds the components in his own workshop and delivers them to General. Arnie's services are an integral part of General's business operations, but Prong B is still satisfied because Arnie performs the services at his own business premises.

Some states that use the ABC test take the position that if a company—for example, a sales firm—has no fixed place of business, a worker cannot satisfy the off-premises test if the services are performed at a temporary work site or where customers or prospective customers are located. This can make it impossible for workers for many types of businesses to qualify as ICs for unemployment compensation purposes.

EXAMPLE: Sam is a home widget installer. He works for a number of different widget sales companies, including Best Buy Widgets. When Best Buy obtains an order, it tells Sam, who goes to the customer's house to install the widget. He does no widget installing at Best Buy's sales office. Sam cannot meet the off-premises test in most states that use the ABC test because he works at temporary work sites where Best Buy's customers are located—their homes. He would be deemed an employee of Best Buy for unemployment compensation purposes.

Prong C: Independent Business or Trade

The final part of the ABC test requires that the worker be engaged in an independently established trade, occupation, profession, or business. This means that the worker's business activity must exist independently of, and apart from, the service relationship with the hiring firm. It must be a stable, lasting enterprise that will survive termination of the relationship with the hiring firm.

There are many ways you can show that a worker is in an independent business or trade. You'll be in good shape if you can prove that the worker does at least some of the following:

- has a separate office or business location
- maintains a business listing in the telephone directory
- owns the equipment needed to perform the services
- employs assistants
- has a financial investment in the business and the ability to incur a loss
- has his or her own liability or workers' compensation insurance
- performs services for more than one unrelated hiring firm at the same time
- is paid by the job rather than by the hour
- possesses all applicable business licenses, and
- files business (Schedule C) federal income tax returns.

Modified ABC Tests

Eight states use a modified version of the ABC test. These states do not require that all three prongs of the standard ABC test discussed above be satisfied; instead, they drop one or two of the three requirements. The list below shows which prongs of the ABC test these states use.

Colorado, Georgia, Idaho, Oregon, Pennsylvania, South Dakota, and Utah drop the B outside-service prong. In these states, a worker will be considered an IC for unemployment compensation purposes if he or she is not under the direction and control of the hiring firm and is engaged in an independent business or trade.

Kansas drops both the B prong and the C independent-business prong. In Kansas, a worker will be an IC for unemployment purposes if he or she is not under the hiring firm's direction and control.

Obviously, it is somewhat easier to establish that a worker is an IC in these states than in those that require all three prongs of the ABC test to be met.

States Using a Modified ABC Test			
State	**Prongs Used**	**State**	**Prongs Used**
Colorado	AC	Oregon	AC
Georgia	AC	Pennsylvania	AC
Idaho	AC	South Dakota	AC
Kansas	A	Utah	AC

The state of Washington uses the ABC test, but adds three additional criteria. These require that, as of the effective date of the contract of service, the worker:

- is responsible for filing a schedule of expenses with the IRS (normally IRS Schedule C)
- has established an account with the Washington Department of Revenue, and
- maintains a separate set of books or records reflecting all income and expenses of his or her business.

Other Tests

Some states use independent means to determine who is and isn't an employee for unemployment compensation purposes.

Montana

Montana essentially uses a modified version of the ABC test, but it requires that a worker obtain an independent contractor exemption certificate from the state Department of Labor and Industry Independent Contractor Central Unit—otherwise, he or she must be covered by unemployment insurance. To obtain the certificate, the applicant must pass the A and B prongs of the ABC test. (Mont. Code § 39-71-417.)

Wisconsin

In Wisconsin, a worker is an IC for UC purposes if he or she:
- holds or has applied for an employer identification number from the IRS or has filed business or self-employment income tax returns with the IRS for the previous year, and
- meets six or more of the following conditions:
  - maintaining a separate business with the worker's own office, equipment, materials, and other facilities
  - operating under contracts to perform specific services for specific amounts of money under which the worker controls the means and method of performing the services
  - incurring the main expenses for the services performed
  - being responsible for the satisfactory completion of the services and liable for failure to satisfactorily complete them
  - being paid solely on a commission, per-job, or competitive-bid basis
  - potentially realizing a profit or suffering a loss
  - having recurring business liabilities or obligations
  - having a business whose success or failure depends on the relationship of business receipts to expenditures

(Wisconsin Statutes § 108.02(12)(b).)

Wyoming

Wyoming looks at three factors to determine whether a worker is an employee for UC purposes. A worker is an IC if he or she:
- is free from control or direction over the details of the performance of services, both by contract and in fact

- represents his or her services to the public as a self-employed individual or an independent contractor, and
- may substitute another individual to perform the services.

(27 Wyoming Statutes § 27-3-104(b).)

Domestic Service

In most states, domestic workers who work in a private home, local college club, or local chapter of a college fraternity or sorority must be covered by unemployment compensation only if they are paid $1,000 or more during any calendar quarter in the current or preceding calendar year. However, some states have different requirements. For example, New York requires that employers provide coverage for domestic workers when their payroll is at least $500 per quarter. Hawaii excludes coverage for domestic workers who work for people with disabilities or on welfare, provided such workers agree in writing to be treated as ICs.

Virginia excludes from unemployment coverage providers of in-home medical services employed by the person receiving the services. In California, recipients of state-provided in-home supportive services must pay unemployment taxes on wages that they or a family member personally pay in order to supplement services compensated by the state.

Agricultural Labor

FUTA's agricultural labor provisions apply to employers who paid wages in cash of $20,000 or more for agricultural labor in any calendar quarter in the current or preceding calendar year, or who employed ten or more workers on at least one day in each of 20 different weeks in the current or immediately preceding calendar year. Most states follow the FUTA provision and, therefore, have limited coverage to services performed on large farms. However, a few states cover services on smaller farms; these states include California, Florida, Minnesota, New York, Rhode Island, Texas, and Washington.

Special Rules for the Construction Industry

Workers in the construction industry have historically been misclassified as independent contractors more often than in most other industries. To help prevent this practice, several states have adopted special unemployment insurance coverage rules for construction workers. These laws make it harder for workers in the construction industry to qualify as ICs and may require prime contractors to cover uninsured workers hired by their subcontractors. In addition, many states enforce these laws by using "stop work orders" that legally prevent an offending hiring firm from continuing to work on (or receive payment for) a construction project until it properly classifies its workers as employees.

California

A person who hires an unlicensed worker or unlicensed subcontractor to perform work requiring a contractor's license is automatically deemed the worker's or subcontractor's employer for all state payroll tax purposes, including unemployment compensation and state income tax. (For details, see the California Employment Development Department's "Construction Industry Information Sheet," at www.edd.ca.gov/pdf_pub_ctr/de231g.pdf.)

Illinois

In Illinois, a person performing services for a construction contractor or subcontractor is presumed to be the hirer's employee for unemployment insurance purposes. Such people qualify as independent contractors only if they:

- pass the ABC test or
- work as sole proprietors or partners in a partnership and pass a special 12-factor test. Starting in 2014, Illinois construction contractors must report to the Illinois Department of Labor various information about all individuals, sole proprietors, or partnerships who receive payments for construction services. (For more information, visit the Illinois Department of Labor website at www.illinois.gov/idol/.)

Maine

Individuals working in construction in Maine are presumed to be employees unless they can pass a special 12-factor test or own construction equipment weighing more than 7,000 pounds. (Maine Rev. Statutes § 105-A.)

Maryland

People working in construction or landscaping services are presumed to be employees in Maryland. A worker overcomes this presumption and qualifies as an IC only if the worker can satisfy a special six-factor test.

Minnesota

Starting September 15, 2012, Minnesota commenced a two-year pilot project requiring all ICs performing construction services in the state who aren't otherwise licensed to register with the Department of Labor and Industry. Additionally, to qualify as ICs, construction workers must pass a nine-factor test. (For details, see the Department of Labor and Industry's website at www.dli.mn.gov/CCLD/register.asp.)

Nebraska

In Nebraska, construction contractors and subcontractors are presumed to be employees of the hiring firm unless they:
- pass the ABC test
- are registered as contractors, and
- are registered to pay unemployment insurance taxes, unless exempt.

New York

In New York, workers in the construction industry are presumed to be employees for unemployment insurance purposes. They qualify as independent contractors only if they:
- pass the ABC test or
- work as sole proprietors or partners in a partnership, a corporation, or another entity and pass a special 12-factor test.

This test is so stringent that few, if any, construction workers can legally qualify as ICs in New York. For details, see www.labor.state.ny.us/ui/dande/fairplayact.shtm.

A somewhat less stringent test went into effect in New York in 2014 for truck drivers. (See the New York State Commercial Goods Transportation Industry Fair Play Act, available at http://open.nysenate.gov/legislation/bill/S5867-2013.)

Pennsylvania

In Pennsylvania, workers in the construction industry are presumed to be employees for unemployment insurance purposes. They qualify as independent contractors only if they:

- have a written IC agreement
- pass the AC test, and
- satisfy a special six-factor test.

For details, see "UCP-32 Employee or Independent Contractor?" available at www.dli.state.pa.us.

Washington

Washington has a special seven-part IC test for construction and electrical contractors. The first three parts comprise the ABC test. In addition, the individual must have a contractor's license and be registered with the Washington Department of Revenue. (For details, see the Washington Employment Security Department website at www.esd.wa.gov/uitax/index.php.)

Statutory Employees

Federal law requires that certain types of workers be covered by unemployment insurance, even if they qualify as ICs. These workers are:

- corporate officers
- drivers who distribute food products, beverages, or laundry, and
- traveling or city salespeople.

These workers are called statutory employees. (See Chapter 3 for a detailed discussion of statutory employees.)

Most states also require that state unemployment compensation be paid for statutory employees. The only exceptions are Maryland, Massachusetts, Montana, and New Hampshire. If you hire a statutory employee in one of these states, you must pay the full 6% FUTA tax, because you won't receive a credit for paying state unemployment taxes.

However, the following states exclude some or all corporate officers from receiving state unemployment insurance: Alaska, California, Delaware, Hawaii, Idaho, Iowa, Michigan, Minnesota, New Jersey, North Dakota, Oklahoma, Oregon, Texas, Washington, and Wisconsin. If you do business in one of these states, check with your state unemployment agency for more information.

Exemptions From Coverage

Most states exempt certain services from unemployment compensation coverage. Thus, hiring firms need not pay for unemployment insurance for workers providing these services, whether or not the workers qualify as ICs under state law.

Services Performed for Relatives

All states exclude services performed for an employer by a spouse or minor child and, with few exceptions, services of an individual in the employ of a son or daughter.

Insurance Agents on Commission

All states (other than California, Iowa, Nevada, New York, and Wyoming) exclude from UC coverage insurance agents who work on commission.

Real Estate Agents on Commission

All states (other than Indiana, Ohio, South Dakota, and West Virginia) exclude from UC coverage real estate agents who work on commission.

Casual Labor

Casual labor is a term used to describe temporary or part-time workers. In most states, workers performing casual labor are not covered by unemployment insurance if the services are not performed in the course of the hiring firm's trade or business.

Casual labor is exempt from UC coverage in all states except Delaware, Idaho, Illinois, Iowa, Maine, Missouri, Montana, Nevada, New Jersey, New Mexico, New York, Oklahoma, South Dakota, Tennessee, Texas, West Virginia, and Wisconsin.

Part-Time Employees of Nonprofits

Most states also exclude from UC coverage part-time services performed for nonprofit organizations exempt from federal tax. This exclusion is limited to remuneration of less than $50 in any calendar quarter in all states, except Alaska (less than $250), Maine (less than $150), and the following states, which have no such exclusion: Delaware, Georgia, Idaho, Iowa, Minnesota, Missouri, Montana, Nevada, New Jersey, New Mexico, New York, Oklahoma, Oregon, Pennsylvania, Tennessee, Texas, Washington, West Virginia, and Wyoming.

Students

The federal unemployment law and all 50 states exclude from unemployment insurance coverage services performed in the employ of a school, college, or university by a student enrolled in and regularly attending classes there. Service by such a student's spouse is also excluded by federal law and most states.

Other Exemptions

Depending on the state, other categories of workers may be exempt from unemployment insurance coverage. Some examples are listed below.

Other Workers Exempt From Unemployment Insurance Coverage: Common Examples

Adjunct college instructors

Licensed barbers and beauticians

Volunteer firefighters

Court reporters

Church employees

Clergy

Elected officials

Election campaign workers

Newspaper delivery workers

Messengers

Student nurses and interns
employed by hospitals

Amateur sports officials

Each state has its own exempt categories. Check with your state unemployment insurance agency for the rules in your state.

State Unemployment Insurance Agencies

You can find a link to your state's unemployment insurance agency at www.servicelocator.org/owslinks.asp.

State Disability Insurance

Five states have disability insurance that provides employees with coverage for injuries or illnesses that are not related to work. These states are California, Hawaii, New Jersey, New York, and Rhode Island. Puerto Rico also has a disability insurance program.

In these states, employees make disability insurance contributions, which are withheld from their paychecks by their employers. Employers must also make contributions in Hawaii, New Jersey, and New York.

The disability insurance coverage requirements are the same as for UC insurance. If you pay UC for a worker, you must withhold and pay disability insurance premiums as well.

Fighting Unemployment Compensation Claims

If a worker you classified as an IC files a claim for unemployment compensation, you don't have to simply accept the worker's assertion that he or she should have been classified as an employee. You have the right to fight the worker's claim, both in administrative proceedings before the state unemployment agency and, if this fails, in state court.

Contesting such a claim can be time-consuming and expensive, but may well be worthwhile because the consequences of having a single worker reclassified as an employee for unemployment insurance purposes can be dire: The business will usually be charged for unpaid unemployment insurance contributions for "all similarly situated" workers and have to pay costly penalties and fines.

Procedures differ from state to state, but are generally handled in the following way.

First, the worker will file a claim with the state UC agency. You'll be notified in writing of the claim and can file a written objection, usually within seven to ten days. Be sure to review your state's worker classification test and timely file an objection explaining why the worker is an IC. Include copies of documentation showing that the worker is in business for himself or herself. You should already have this material in your files. (See Chapter 11.)

Next, the UC agency will determine whether the worker is eligible to receive UC benefits. There's usually no hearing at this stage.

If you don't or the worker doesn't like the UC agency's ruling, either of you can demand a hearing. This is usually held at the UC agency's office before a hearing officer (called a referee in many states) on the agency's staff. You should present your written documentation showing that the worker was an IC, not your employee. You can testify and also present oral testimony from other witnesses—for example, supervisors who dealt with the worker. The more relevant, persuasive evidence you have, the better off you'll be.

Before the hearing, ask to see the UC agency's complete file on the claim; it may contain inaccurate statements you'll need to refute.

You're entitled to have an attorney represent you at the hearing. If you can afford it, this is not a bad idea, because an adverse ruling may result in audits of other workers you've classified as ICs. And the IRS or other agencies may decide to audit as well.

Fighting Unemployment Compensation Claims (continued)

Either side can then appeal the UC hearing officer's ruling to a state administrative agency, such as a board of review. You should usually hire a lawyer to do this. Such appeals are usually not successful.

Finally, you can appeal to your state courts. Again, you will probably need the help of a lawyer to do this. Your appeal will likely fail unless you can show that the prior rulings were contrary to law or not supported by substantial evidence.

In California, New Jersey, and Rhode Island, disability insurance is handled by the state unemployment compensation agency. The same employee records are used for UC and disability—and employers submit contribution reports for both taxes at the same time. New York's disability program is administered by its Workers' Compensation Board. In Hawaii, the Temporary Disability Insurance Division of the Department of Labor and Industrial Relations handles disability insurance.

California also provides paid family leave insurance for workers who are covered by the state disability insurance program. This program lets employees take up to six weeks of partially paid time off to care for a seriously ill child, spouse, parent, or domestic partner, or to bond with a new minor child. The insurance is funded by contributions from employees' wages; employers must withhold this money and send it in to the state, although they don't have to contribute their own money to the fund.

State Income Taxes

All states except Alaska, Florida, Nevada, South Dakota, Texas, Washington, and Wyoming have income taxation. New Hampshire and Tennessee impose income taxes only on interest and dividend income. If you do business in a state that imposes state income taxes, you must withhold the applicable tax from your employees' paychecks and pay it over to the state taxing authority. No state income tax withholding is required for workers who qualify as ICs.

It's very easy to determine whether you need to withhold state income taxes for a worker: If you are withholding federal income taxes, then you must withhold state income taxes as well. Contact your state tax department for the appropriate forms.

If a worker qualifies as an IC for IRS purposes, you won't need to withhold federal or state income taxes.

State Tax Offices

You can find a link to your state's income tax office at www.taxadmin.org/fta/link/default.php.

Workers' Compensation

Basics of the Workers' Compensation System.. 156

    ICs Are Excluded 157

    Cost of Coverage 159

Who Must Be Covered 160

Exclusions From Coverage 160

    States With Employee Minimums 160

    Casual Labor 161

    Domestic Workers 162

    Sole Proprietors, Partners, LLCs, and Corporate Officers 164

    Farm Labor 164

    Undocumented Workers 164

    Other Exemptions 165

Classifying Workers for Workers' Compensation Purposes 165

    Common Law States 167

    Relative-Nature-of-the-Work Test 170

    Other Tests 173

    Special Rules for the Construction Industry 176

    Consequences of Misclassifying Workers 177

If Your Workers Are ICs 178

    Require ICs to Get Their Own Coverage 178

    Coverage for an IC's Employees 179

Obtaining Coverage 181

T his chapter provides an overview of the workers' compensation system. It also explains when you need to provide workers' compensation insurance for workers and what happens if you don't. Each state has its own workers' compensation laws and insurance system. No two states' laws are exactly alike, including when it comes to how they define who is and is not an independent contractor. You should check with your state workers' compensation agency to find out the details of your state's requirements. Links to all 50 state workers' compensation agency websites are at: www.dol.gov/owcp/dfec/regs/compliance/wc.htm.

Basics of the Workers' Compensation System

Each state's workers' compensation system is designed to provide replacement income and pay medical expenses for employees who suffer work-related injuries or illnesses. Benefits may also extend to the survivors of workers who are killed on the job. To pay for this, employers in all but one state—Texas, where workers' compensation is optional—are required to pay for workers' compensation insurance for their employees, either through a state fund or a private insurance company. Employees do not pay for workers' compensation insurance.

Before the first workers' compensation laws were adopted, about 100 years ago, an employee injured on the job had only one recourse: Sue the employer in court for negligence—a difficult, time-consuming, and expensive process. The workers' compensation laws changed this by establishing a no-fault system. Although employees can't sue in court, they are entitled to receive compensation without having to prove that the employer caused the injury. In exchange for paying for workers' compensation insurance, employers are spared from having to defend against lawsuits by injured employees and paying out damages.

An employee who is injured on the job can file a workers' compensation claim and collect benefits from the employer's workers' compensation insurer, but cannot sue the employer in court except in rare cases where the employer intended to cause the injury—for example, by beating up an employee. Workers' compensation benefits are set by state law and are usually modest. Employees can obtain reimbursement for medical and rehabilitation expenses and lost income, but can't collect benefits for pain and suffering or mental anguish caused by an injury.

> **EXAMPLE:** Sam, a construction worker for the Acme Building Company, accidentally severs a muscle in his arm while using a power saw on an Acme construction site. Sam is an Acme employee, so he is covered by Acme's workers' compensation insurance policy.
>
> Sam may file a workers' compensation claim and receive benefits from Acme's workers' compensation insurer. Under the law of Sam's state, Sam is entitled to receive a maximum of $2,520 to make up for lost income plus medical and rehabilitation expenses. All of this is paid by Acme's workers' compensation insurer, not by Acme itself. This is all Sam is entitled to collect. He cannot sue Acme in court for damages arising from his injuries.

Employers' workers' compensation premiums will likely go up if many employees file workers' compensation claims, but the premiums will almost certainly be cheaper than defending against employee lawsuits. Indeed, many employers purchase workers' compensation insurance even if they aren't required to do so. State laws typically allow these exempt employers to "opt in" to the workers' comp system. This allows an employer to ensure that its employees are compensated for workplace injuries—and that they can't file a lawsuit against the employer.

ICs Are Excluded

You generally need not provide workers' compensation for a worker who qualifies as an IC under your state workers' compensation law. (See below.) This can result in substantial savings. However, unlike employees who are covered by workers' compensation, ICs can sue you for work-related injuries if your negligence—that is, carelessness or failure to take proper safety precautions—caused or contributed to the injury. You will also be held responsible for the negligence of your employees.

EXAMPLE: Trish, a self-employed trucker, contracts to haul produce for the Acme Produce Company. Trish is an IC, and Acme does not provide her with workers' compensation insurance. Trish loses her little finger when an Acme employee negligently drops a load of asparagus on her hand. Trish is an IC, so she can't collect workers' compensation benefits from Acme's insurer, but she can sue Acme in court for negligence. If she can prove Acme was negligent, Trish can collect damages not only for her lost wages and medical expenses, but for her pain and suffering as well.

These damages could far exceed the modest sums that workers' compensation benefits would have provided. Had Trish been Acme's employee, it might have cost Acme several hundred dollars a year to provide workers' compensation coverage for her. But because she was an IC, it could cost Acme tens of thousands of dollars to defend her lawsuit and pay out damages.

Of course, an IC must actually prove that your negligence or that of your employees caused or contributed to the work-related accident in order to recover any damages at all. So, if you and your employees weren't negligent, the IC may end up losing a lawsuit or decide not to file one in the first place.

As an alternative to filing a risky negligence claim, an injured worker you classified as an IC may file a workers' comp claim alleging that he or she should have been classified as an employee. In some states, such claims are filed against a special uninsured employers fund (for example, California has established an Uninsured Employers Benefits Trust Fund for this purpose). Such a claim might be also be filed on the injured worker's behalf by a third party legally allowed to enforce the worker's rights, such as his or her health insurer, a governmental benefit program, or your general liability insurer.

Protect Yourself With Liability Insurance

If you hire ICs, you must obtain general liability insurance to protect yourself against personal injury claims by people who are not your employees. A general liability insurer will defend you in court if an IC, a customer, or any other nonemployee claims you caused or helped cause an injury. The insurer will also pay out damages or settlements up to the policy limits. Such insurance can be cheaper and easier to obtain than workers' compensation coverage for employees, so you can still save on insurance premiums by hiring ICs rather than employees.

Cost of Coverage

The cost of workers' compensation varies from state to state and depends upon a number of factors, including:

- how generous the state's workers' comp benefits are (each state's benefits differ)
- the size of the employer's payroll
- the nature of the industry, and
- how many claims have been filed in the past by the employer's employees.

The national median for premium rates in 2010 was $2.04 per $100 of payroll. Premium rates ranged from a low of $1.02 in North Dakota to a high of $3.33 in Montana.

As you might expect, it costs far more to insure employees in hazardous occupations such as construction than it does to provide coverage for those in relatively safe jobs such as clerical work. It might cost $500 to $600 a year to insure a clerical worker and perhaps ten times as much to insure a roofer or lumberjack.

Depending on the laws of the state where you do business, you can obtain workers' compensation insurance from a state fund, a private insurer, or both.

Who Must Be Covered

You must use a two-step analysis to determine whether you must provide a worker with workers' compensation insurance.

First, determine whether workers' compensation coverage is necessary if the worker qualifies as an employee under your state's workers' compensation law. Most states exclude certain types of workers from workers' compensation coverage. (See below.) You won't have to provide coverage for workers who fall within these exclusions. Some states also don't require coverage unless you have a minimum number of employees.

Second, if there is no exclusion for the workers involved, you must determine whether the workers should be classified as employees or ICs under your state's workers' compensation law. If they're employees, you'll have to provide coverage; if they're ICs, you won't. States use different tests to classify workers for workers' compensation purposes; these rules are covered below.

Exclusions From Coverage

Most states exclude certain types of workers from workers' compensation coverage. The nature and scope of these exclusions vary somewhat from state to state. Check the workers' compensation law of your state—or ask your workers' compensation carrier to do so—to see how these exclusions operate in your state.

States With Employee Minimums

The workers' compensation laws of several states exclude employers having fewer than a designated number of employees. In other words, if you have fewer than this number of employees, you don't need to obtain workers' compensation insurance for anyone—employees or ICs.

State Requirements for Workers' Compensation Coverage

State	Employees Required	State	Employees Required
Alabama	Five or more	Missouri	Five or more
Arkansas	Three or more	New Mexico	Three or more
Florida	Four or more (one or more for construction trades)	North Carolina	Three or more
		South Carolina	Four or more
Georgia	Three or more	Tennessee	Five or more
Michigan	Three or more	Virginia	Three or more
Mississippi	Five or more	Wisconsin	Three or more

Casual Labor

Most states exempt casual workers from workers' compensation coverage. Who qualifies as a casual worker varies from state to state. In most states, casual labor must be for a brief time period, for work outside the hiring firm's usual course of business. These states include Alabama, Arizona, Arkansas, Connecticut, Delaware, Florida, Georgia, Idaho, Indiana, Iowa, Kentucky, Minnesota, Montana, Nevada, New Jersey, New Mexico, North Carolina, North Dakota, Ohio, Oregon, Pennsylvania, South Carolina, Tennessee, Utah, Vermont, Virginia, West Virginia, Wisconsin, and Wyoming.

> EXAMPLE: The Acme Widget Company hires Sue, a caterer, to cater a retirement dinner for Acme's president. Catering is outside Acme's usual course of business—manufacturing widgets. And Sue is hired for a temporary period—to plan a single event. Sue would qualify as a casual laborer under most state laws and would not have to be covered by workers' compensation.

This exception is narrow. For example, people hired to do temporary maintenance or repair work would generally not be casual laborers because such work is usually considered part of a firm's normal course of business.

> EXAMPLE: A partition broke at a soft drink distributor's place of business, smashing hundreds of bottles. The firm hired two workers for a single day to clean up the bottles. The workers were not casual laborers because such clean-up work was a regular part of a soft drink distributor's business; such breakage was an inherent risk of the business. (*Graham v. Green*, 156 A.2d 241 (1959).)

Some states have more liberal rules. To find out about your state's law, contact your state workers' compensation office.

Domestic Workers

The states listed in the chart below require that domestic workers who work in private homes be covered by workers' compensation if their salary or time worked exceeds a threshold amount. States not listed in the chart exclude these domestic employees from workers' compensation coverage. Domestic or household workers include housekeepers, gardeners, babysitters, and chauffeurs. (See Chapter 7 for more information about dealing with these types of workers.)

Workers' Compensation Requirements for Domestic Workers	
State	**Domestic Workers Who Must Be Covered**
Alaska	All, except part-time babysitters, cleaners, and similar part-time help
California	All who, during 90-day period preceding injury, work 52 or more hours or earn more than $100; workers employed by a parent, spouse, or child are excluded
Colorado	All who work more than 40 hours per week for five or more days per week for one employer
Connecticut	All who work more than 26 hours per week for one employer
Delaware	All paid $750 or more in any three-month period by single private home

Workers' Compensation Requirements for Domestic Workers (continued)

State	Domestic Workers Who Must Be Covered
Hawaii	All whose wages are $225 or more during the current calendar quarter and during each completed calendar quarter of the preceding 12-month period
Illinois	All who work 40 or more hours per week for 13 or more weeks during a calendar year for any household or residence
Iowa	All who earn $1,500 or more during the 12 months before an injury
Kansas	All, if employer had a $20,000 or more gross payroll for all employees (domestic and nondomestic) during the preceding year
Kentucky	All, if two or more work in a private home 40 or more hours per week
Maryland	All paid $1,000 or more in any calendar quarter by single private home
Massachusetts	All who work 16 or more hours per week for one employer
Minnesota	All who earn $1,000 or more in any three-month period or $1,000 or more in any three-month period during the previous year from the same private household
New Hampshire	All
New Jersey	All except casual labor
New York	All employed by the same employer for 40 or more hours per week
Ohio	All who earn $160 or more in any calendar quarter from one employer
Oklahoma	All, if employer had a $10,000 or more gross payroll for all domestic employees during the preceding year
South Carolina	All, if four or more domestic workers are employed, except employers whose total annual payroll the previous calendar year was less than $3,000
South Dakota	All employed 20 or more hours in any calendar week and for more than six weeks in any 13-week period
Utah	All regularly employed for 40 or more hours per week by the same employer
Washington	All, if two or more domestic workers are regularly employed in a private home 40 or more hours per week

Sole Proprietors, Partners, LLCs, and Corporate Officers

In all states, sole proprietors, partners in partnerships, and limited liability company members aren't required to purchase workers compensation unless and until they have employees who aren't owners. However, most states will allow them to cover themselves for workers comp if they choose to.

The same rule holds true in most states for officers of small "closely held" corporations with no employees. However, some states require corporate officers to own at least a specified percentage of the corporate stock to be exempt—for example, Idaho, Louisiana, and Michigan require less-than-10% shareholders to be covered.

In some states, sole proprietors, partners, LLC members, or corporate officers must file a document with their state workers' compensation agency to obtain their exemption from their state's workers' comp requirements.

Farm Labor

Most states exempt all or some farm laborers from workers' compensation coverage. These include: Arkansas, Colorado, Florida, Georgia, Illinois, Indiana, Iowa, Kansas, Kentucky, Louisiana, Maine, Maryland, Minnesota, Mississippi, Missouri, Nebraska, Nevada, New Mexico, New York, North Carolina, North Dakota, Oklahoma, Pennsylvania, Rhode Island, South Carolina, South Dakota, Tennessee, Utah, Vermont, Virginia, and West Virginia.

However, not every person who works on a farm is exempt. For example, a horse trainer isn't considered a farm worker when it comes to eligibility for workers' compensation benefits.

Check with your state workers' compensation agency for details on your state's farm labor exemption.

Undocumented Workers

Some states—including Arizona, California, Florida, Montana, Nevada, New York, Texas, and Utah—cover undocumented workers in their workers' compensation statutes. Other states, such as Idaho and Wyoming, expressly exclude undocumented workers. Still other states, like Colorado, have not yet dealt with the issue, leaving it to the courts to struggle with whether to cover these workers.

Other Exemptions

Depending on the state, other categories of workers may be exempt from workers' compensation coverage requirements. These may include, for example:

- business owners' immediate family members, including parents, spouses, and children
- licensed real estate brokers and salespersons who work on commission
- licensed construction contractors
- employees of charities
- volunteer workers
- owner-operator truck drivers
- newspaper delivery workers
- church employees and clergy
- professional athletes
- insurance salespeople
- amateur sports officials
- youth sports league participants, and even
- employer-sponsored bowling teams.

Check your with your state workers' compensation agency for details.

Classifying Workers for Workers' Compensation Purposes

If none of the exclusions discussed above applies, you must determine whether a worker is an employee or IC under your state's workers' compensation law. If the worker qualifies as an employee, you must provide workers' compensation coverage. (Remember: Even if you know what a worker's classification is under the IRS rules and under your state's unemployment laws, you must still check your state's workers' compensation rules, because a worker may be an IC under some rules but an employee under others.) Each state has its own workers' compensation law with its own definition of who is an employee. However, these state laws follow one of three patterns:

- Most states classify workers for workers' compensation purposes using the common law right-of-control test.
- Other states use a relative-nature-of-the-work test, either alone or in conjunction with the common law test.
- A few states use different classification schemes.

Find your state in the chart below and read the applicable discussion. For more detailed information on your state's workers' compensation laws, contact the state workers' compensation agency.

When reading about your state's law, keep in mind that most state workers' compensation agencies and state courts interpret their workers' compensation laws to require coverage. Workers' compensation laws are designed to help injured workers, and it's considered beneficial for society as a whole to have as many workers as possible covered. Indeed, in several states, all workers are legally presumed to be employees for workers' compensation purposes and it's up to the hiring firm to prove that they are ICs. These states include Arizona, California, Colorado, Connecticut, Delaware, Hawaii, Massachusetts, New Hampshire, New Jersey, New Mexico, North Dakota, Wisconsin, and Washington.

Generally, if there is any uncertainty as to how a worker should be classified for workers' compensation purposes, state workers' compensation agencies and courts find that the worker is an employee who should be covered.

Certification of Worker IC Status

In several states, all ICs, or just ICs in certain occupations, can obtain or file a document certifying that they are ICs for state workers' compensation purposes. These states include: Arkansas, Colorado, Indiana, Maine, Minnesota, Montana, North Dakota, Oklahoma, Rhode Island, and Texas. Depending on the state involved, the worker may simply have to sign a document certifying that he or she is free of control.

In some states, such as Montana, Maine, and North Dakota, the state workers' comp agency conducts an investigation to make a finding that the worker qualifies as an IC under the state's law.

The legal weight such certification carries varies from state to state. In some states, it only establishes a rebuttable presumption that the worker is an IC. In others, it provides the hiring firm with more protection against workers' comp claims.

Common Law States

A majority of state workers' compensation statutes define an employee as one who works for, and under the control of, another person for hire. The right to control the details of the work is the primary test used to determine whether an employment relationship exists. (See Chapter 2 for a detailed discussion of the common law test.)

The list of factors used to measure control varies somewhat from state to state. But the goal of the test is the same: to determine whether the hiring firm has the right to direct and control how the worker does the job, including when, where, and how the work is to be done.

The following states use a four-factor test to determine whether a hiring firm has the right of control: Alabama, Idaho, Kansas, Mississippi, Montana, New Jersey, North Dakota, Oregon, South Carolina, and West Virginia. These states ask whether:

- the hiring firm has the right to control the performance of the work itself, including how, when, and where it is performed

- the hiring firm has the right to discharge the worker and the worker has the right to quit at any time
- the worker is paid by the job or on a time basis (hourly, weekly, or monthly), and
- the hiring firm supplies the worker with valuable equipment.

These factors are covered in detail in Chapter 2.

It is far harder to prove that a person is an IC than an employee for workers' compensation purposes. Because the states want as many workers as possible to qualify as employees, a worker can be deemed an employee for workers' compensation purposes even if only one or two of these factors point to employee status. For example, a worker's compensation auditor who discovers that you have the right to fire a worker at any time for any reason, or for no reason at all, will likely stop the audit right there and conclude that the worker is your employee. But even workers who cannot be discharged without cause may be employees.

> **EXAMPLE:** Joanna, a driver who delivers dry cleaning for Ace Cleaners, designates her own delivery schedules and routes and is otherwise not controlled by Ace. She has a one-year contract with Ace that provides that she cannot be terminated or quit unless there is a material breach of the contract.
>
> However, Ace provides her with a delivery truck and pays her by the hour. These two factors alone are enough to show an employment relationship for workers' compensation purposes. The combination of a hiring firm providing a driver with a vehicle and paying by the hour is almost always enough to show employment for workers' compensation purposes.

To establish IC status, you will usually have to satisfy all the factors discussed above, with the possible exception of the method of payment.

> **EXAMPLE:** An Oregon country club hired Marcum, an unemployed logger, to prune dead wood from some trees on the golf course. He was injured after a few days on the job and filed a workers' compensation claim alleging that he was the club's employee. The court concluded that Marcum was an IC. The court examined all

State Tests Used to Classify Workers

State	Test	State	Test
Alabama	Common law	Mississippi	Common law and relative nature of work
Alaska	Relative nature of work	Montana	Other (certification procedure)
Arizona	Common law		
Arkansas	Common law and relative nature of work	Nebraska	Common law
California	Common law and other (economic reality)	Nevada	Common law
		New Hampshire	Common law
Colorado	Common law	New Jersey	Common law and relative nature of work
Connecticut	Common law		
Delaware	Common law	New Mexico	Common law
Dist. of Columbia	Common law	New York	Common law and relative nature of work
Florida	Common law		
Georgia	Common law	North Carolina	Common law
Hawaii	Common law and relative nature of work	North Dakota	Common law and relative nature of work
Idaho	Common law	Ohio	Common law
Illinois	Common law	Oklahoma	Common law
Indiana	Common law	Oregon	Common law
Iowa	Common law and relative nature of work	Pennsylvania	Common law
		Rhode Island	Common law
Kansas	Common law	South Carolina	Common Law
Kentucky	Common law	South Dakota	Common law
Louisiana	Common law and relative nature of work	Tennessee	Common law
Maine	Common law	Texas	Common law
Maryland	Common law and relative nature of work	Utah	Common law
		Vermont	Common law
Massachusetts	Common law	Virginia	Common law
Michigan	Other	Washington	Other (6 factors)
Minnesota	Common law and other	West Virginia	Common law
Missouri	Common law and relative nature of work	Wisconsin	Other (9 factors)
		Wyoming	Common law

four factors and found that three indicated IC status, while the fourth was neutral:

- **Control.** There was no direct evidence of control over Marcum by the club. A club member testified that he simply told Marcum which trees to prune. He did not tell him how to do the work or what hours to work. Marcum hired and paid his own assistant. Marcum had no continuing relationship with the club—he had never worked for it before.
- **Right to fire.** The club member who hired Marcum stated that he felt he could terminate Marcum's contract only if he was not properly doing the job.
- **Equipment.** Marcum furnished all his own equipment, including saws and a pickup truck.
- **Method of payment.** Marcum was paid $25 per tree. Such piecework payment indicated neither IC nor employee status.

(*Marcum v. State Accident Ins. Fund,* 565 P.2d 399 (1977).)

Relative-Nature-of-the-Work Test

Although most states now use the common law test discussed above, there is a growing trend to use a test that makes it more difficult to establish that a worker is an IC. Sometimes called the "relative nature of the work" test, this test is based on the simple notion that the cost of industrial accidents should be borne by consumers as part of the cost of a product or service. If a worker's services are a regular part of the cost of producing your product or service, and the worker is not running an independent business, this test dictates that you should provide workers' compensation insurance for the worker and then, if necessary, pass the cost on to the ultimate consumer of your product or service.

On the other hand, you don't need to provide workers' compensation if a worker is running an independent business and the worker's services aren't a normal, everyday part of your business operations, the cost of which you regularly pass along to your customers or clients.

To determine whether a worker is an employee or IC under the relative-nature-of-the-work test, you must answer the following question: Is the worker running an independent business?

If the answer to this question is "no," then the worker is an employee for workers' compensation purposes. If the answer to this question is "yes," then you still need to ask another question to determine the worker's status: Are the worker's services a regular part of the cost of your product or service?

If the answer to this second question is "yes," then the worker is an employee. Only if the answer to this question is "no" is the worker an IC for workers' compensation purposes.

Many states use the relative-nature-of-the-work test in conjunction with the common law control test. If a worker's status is unclear under the common law test, these states use the relative-nature-of-the-work test. Other states use only the relative-nature-of-the-work test. Check the chart, "State Tests Used to Classify Workers," above, to find out what your state does.

Independent Business

If a worker is running an independent business and has the resources to provide his or her own insurance coverage for work-related accidents, it's reasonable not to require the worker's clients or customers (meaning you, the employer) to provide such coverage.

It's much easier for highly skilled (and well-paid) workers to qualify as ICs under this test. Highly skilled workers are more likely to have their own independent businesses and to earn enough to be financially responsible for their own accidents.

To decide whether a worker has an independent business, workers' compensation agencies and courts will probably examine many of the factors from the common law test. (See Chapter 2 for a complete discussion of the common law test.) A worker is more likely to be viewed as operating an independent business if he or she:

- makes his or her services available to the public—for example, by advertising
- does not work full time for you
- has multiple clients and income sources, and
- has the right to reject jobs you offer.

Part of the Cost of Your Product or Service

A worker's services will likely be viewed as a regular part of the cost of your product or service if:

- the worker's services are a regular part of your company's daily business operations
- the work is continuous rather than intermittent, and
- the worker provides continuous services rather than contracting for a particular job.

EXAMPLE 1: Ceradsky was killed while operating a milk truck owned by Purcell. Purcell had contracted with a cheese manufacturer to pick up milk from farmers along a specific route and deliver it to the cheese factory. Although Purcell and Ceradsky had been classified as ICs by the cheese company, Ceradsky's survivors applied for workers' compensation benefits from the company. The court held that they were entitled to benefits because both Ceradsky and Purcell were employees of the cheese company under the relative nature of the work test.

The milk hauling was not an independent business, but only an aid to the cheese company's production process. Purcell and Ceradsky hauled milk exclusively for the cheese company six days a week for years. They didn't earn enough money from the work to be expected to purchase their own insurance. In addition, the milk hauling work was a regular and essential part of the company's cheese production process, the costs of which were passed on to cheese consumers. (*Ceradsky v. Mid-America Dairymen, Inc.*, 583 S.W.2d 193 (Mo. App. 1979).)

EXAMPLE 2: Ostrem suffered an eye injury while installing a diesel engine in a piece of heavy equipment owned by a construction company. He applied for workers' compensation benefits from the company and was denied them because he was an IC under the relative-nature-of-the-work test. First, the court found that he operated an independent business: He was highly skilled, established his own rate of pay and hours, took out a business license, was generally unsupervised, had multiple clients, and made his services available to the public. Second, Ostrem's work wasn't part of the construction company's regular business. He had been hired to install one engine only and had never worked for the company before. The job should have taken only a few days. He had been hired to complete a single job, not to do regular work of the construction company. (*Ostrem v. Alaska Workmen's Compensation*, 511 P.2d 1061 (Alaska 1973).)

Other Tests

A few states have somewhat different tests for determining who qualifies as an employee for workers' compensation purposes. The tests used in Michigan, Minnesota, Washington, and Wisconsin define who is an employee much more clearly than the common law or relative-nature-of-the-work tests. If you're doing business in one of these states, consider yourself fortunate. California is another story, however.

California

California uses at least two tests—the common law test and a second, economic reality test. If a worker is an IC under the common law test, the California Workers' Compensation Appeal Board and courts will use the second, broader economic reality test to try to find employee status for the worker involved.

Under the California version of the common law control test, a worker is more likely to be viewed as an IC if the worker:

- has the right to control the manner and performance of the work
- has a monetary investment in the work
- controls when the work begins and ends
- supplies tools and instruments needed for the work
- has a license to perform the work
- is paid by the project rather than by unit of time (such as hourly payment)
- is engaged in a distinct occupation or business
- is highly skilled
- can't quit at any time

- was hired for a temporary and fixed, rather than indefinite, time, and
- believes, along with the hiring firm, that the relationship is an IC relationship.

California uses the same economic reality test that applies to federal labor laws. (See Chapter 8 for a discussion of how to determine if someone is an IC under federal labor laws.) This test emphasizes whether the worker functions as an independent business or is economically dependent upon the hiring firm. Under this test, a worker is more likely to be viewed as an IC if:

- the worker has the right to control the manner and performance of the work
- the worker's opportunity for profit or loss depends on his or her managerial skill
- the worker supplies tools and instruments needed for the work
- the services rendered require a special skill
- the working relationship is temporary rather than permanent, and
- the services rendered are not an integral part of the hiring firm's business.

In addition, California has a few special rules. It requires all roofing companies to have workers' compensation coverage, even if they have no employees. Real estate brokers are required to provide workers' compensation coverage for their salespeople, even though they are usually classified as ICs under federal tax law.

Michigan

Under Michigan law, workers are employees for workers' compensation purposes when they:

- do not maintain a separate business,
- do not hold themselves out to and render services to the public, and
- do not employ other workers. (Michigan Workers' Disability Compensation Act, Mich. Comp. Laws § 161(1)(n).)

Minnesota

Minnesota uses the four-factor common law test described above, but has designed special rules that are intended to create a safe harbor for 34 specific occupations—that is, workers who come within the rules are deemed ICs. These occupations include: artisans, barbers, bookkeepers and accountants, bulk oil plant operators, collectors, consultants, domestic workers, babysitters, industrial home workers, laborers, orchestra musicians, several types of salespeople or manufacturer's representatives, agent drivers, photographers, models, some professional workers, medical doctors providing part-time services to industrial firms, real estate and securities salespeople, registered and practical nurses, unlicensed nurses, taxicab drivers, timber fellers, buckers, skidders and processors, sawmill operators, truck owner-drivers, waste materials haulers, messengers and couriers, variety entertainers, sports officials, jockeys, and trainers. (See Minn. R. § 5224 and following.) Individuals who work in the construction industry must pass a special nine-point test to qualify as ICs.

Washington

In Washington, a worker is an IC for workers' compensation purposes if all of the following six conditions are met:
- The worker is free from control while performing the services.
- The worker's services are either:
  - outside the hiring firm's usual course of business
  - performed outside the firm's places of business, or
  - performed at a workplace for which the worker pays.
- The worker is engaged in an independent business, or has a principal place of business that is eligible for a federal income tax business deduction.
- The worker is responsible for filing a Schedule C or similar form with the IRS listing the worker's business expenses.
- The worker pays all applicable state business taxes, obtains any necessary state registrations, and opens an account with the State Department of Revenue.
- The worker maintains a separate set of books or records showing all income and expenses of his or her business.

(Wash. Rev. Code § 51.08.195.)

For more information, see "Independent Contractor Guide: A Step-by-Step Guide to Hiring Independent Contractors in Washington State," available at www.lni.wa.gov. Click "Get a Form or Publication," then enter "independent contractor" in the search field.

Wisconsin

Under Wisconsin's test, a worker is an IC for workers' compensation purposes if he or she satisfies nine conditions. He or she must:

- maintain a separate business with his or her own office, equipment, materials, and other facilities
- hold or have applied for a federal employer identification number
- operate or contract to perform specific services or work for specific amounts of money, with the worker controlling the means of performing the services or work
- incur the main expenses related to the service or work that he or she performs under contract
- be responsible for completing the work or services and be liable for failing to complete it
- receive compensation for work or services performed under a contract on a commission, per job, or competitive bid basis and not on any other basis—for example, hourly payment
- be able to realize a profit or suffer a loss under the contracts
- have continuing or recurring business liabilities or obligations, and
- have a business setup in which success or failure depends on the relationship of business receipts to expenditures.

(Wis. Stat. § 102.07(8).)

Special Rules for the Construction Industry

Historically, the construction industry has had some of the highest levels of worker misclassification. To combat this, in recent years, several states have adopted laws making it harder to classify workers in the construction industry as ICs for purposes of state workers' compensation and unemployment insurance laws. These states include California, Illinois, Maryland, New York, and Pennsylvania. See the discussion in Chapter 5 for details.

Consequences of Misclassifying Workers

You will suffer harsh penalties if you misclassify an employee as an IC for workers' compensation purposes and have no workers' compensation insurance.

Most state workers' compensation agencies maintain a special fund to pay benefits to injured employees whose employers failed to insure them. You will be required to reimburse this fund or pay penalties to replenish it.

In addition, in most states, the injured worker can sue you in court for personal injuries. Most states try to make it as easy as possible for injured employees to win such lawsuits by not allowing you to raise legal defenses you might otherwise have, such as that the injury was caused by the employee's own carelessness.

You will also have to pay fines imposed by your state workers' compensation agency for your failure to insure. These fines vary widely, ranging from $250 to $5,000 per employee. The workers' compensation agency may also obtain an injunction—a legal order—preventing you from doing business in the state until you obtain workers' compensation insurance.

If you are doing business as a sole proprietor or partnership, you will be personally liable for these damages and fines. And the fact that your business may be a corporation won't necessarily shield you from personal liability. In some states, shareholders of an uninsured corporation may be personally liable for the injuries sustained by the corporation's employees. For example, in California, any shareholder of an illegally uninsured corporation who holds 15% or more of the corporate stock, or at least a 15% interest in the corporation, may be held personally liable for the resulting damages and fines.

Finally, in almost all states, failure to provide employees with workers' compensation insurance is a crime—a misdemeanor or even a felony. An uninsured employer may face criminal prosecution, fines, and, in rare cases, prison.

If Your Workers Are ICs

If you decide, after reviewing your state's test and your circumstances, that your workers can properly be classified as ICs, then you don't have to purchase workers' compensation coverage for them. To make sure that you don't run into problems with your insurance company, however, you may have to require your ICs to get their own workers' compensation coverage. And, depending on your situation, you may have to purchase workers' compensation insurance for employees of your ICs, even though you don't have to cover the ICs themselves.

Require ICs to Get Their Own Coverage

Many companies require workers classified as ICs to purchase workers' compensation coverage for themselves. Such coverage is available in most states, even if the IC is running a one-person business.

"If Any" Workers' Compensation Policies

In most states, special low-cost workers' compensation insurance policies are available to independent contractors who have no employees. These are often called "if any" policies because they provide coverage for work-related injuries during the policy period "to the policyholder's employees, if any." In other words, they provide no coverage for the IC, only employees he or she may hire.

Such an "if any" policy will protect a firm that hires the IC (and its own workers' comp insurer) from claims brought against the IC by someone alleging to have been injured while working for him or her. Moreover, if the IC does hire an employee, the policy satisfies the requirement that the employer have workers' compensation insurance, thus avoiding noncomplying employer fines.

You might be wondering why you should care whether your IC has workers' compensation insurance. The answer is that your own insurance carrier might require it. If an IC doesn't have his or her own workers' compensation coverage, there is a risk that the IC might get injured and then claim to be your employee, just to get workers' compensation benefits. That's why many insurance carriers will require you to include an IC who doesn't have his or her own workers' compensation coverage on your own policy (and to pay additional premiums for the privilege). Most insurers audit payroll and other employment records at least once a year to make sure companies are adequately covered and are paying sufficient premiums.

If you require your ICs to carry their own insurance, ask them to give you a certificate of insurance—a document issued by the insurer that provides written proof that the IC has a policy. The certificate should state the insurer, insurance agency, type of coverage, policy number, effective dates, limits and exclusions, certificate holders, and any special provisions. And make sure that the policy is in effect during the period when the IC will be working for you. Keep the certificate in your files so you can show it to your workers' compensation insurer when you are audited.

> **RESOURCE**
>
> **Need more information on workers' compensation audits?** You can find detailed guidance in *Worker's Compensation: A Field Guide for Employers*, by Ed Priz and Scott Priz (Advanced Insurance Management LLC), and *Managing Workers' Compensation: A Guide to Injury Reduction and Effective Claim Management*, by Keith R. Wertz and James J. Bryant (Lewis Publishers).

Even if your ICs have their own workers' compensation insurance, you'll still need liability insurance because an IC can still sue you for injuries caused by your own carelessness. The workers' compensation provisions that prevent injured employees from filing lawsuits won't protect you because you are not the IC's employer. You might even face a lawsuit from an IC's workers' compensation carrier, seeking to get back the money it laid out to pay the IC's claim.

Coverage for an IC's Employees

In every state except Alabama, California, Delaware, Iowa, Maine, Rhode Island, and West Virginia, you may have to provide workers' compensation benefits for the employees of ICs you hire, depending on the circumstances.

Under the laws of most states, an IC's uninsured employees are considered to be your employees for purposes of workers' compensation insurance if:

- the IC doesn't get workers' compensation for them, and
- they perform work that is part of your regular business—that is, work customarily done in your business and similar businesses.

The purpose of these laws is to discourage employers from subcontracting work out to uninsured ICs in order to avoid the costs of workers' compensation coverage.

> EXAMPLE: The Diamond Development Company, a residential real estate developer, is building a housing subdivision. It hires Tom, a painting subcontractor, to paint the houses. Tom is an IC who has sole control over the painting work. Tom hires 40 painters to do the work for him. Although they are all Tom's employees, Tom doesn't get workers' compensation coverage for them.
>
> Andy, one of Tom's employees, is injured on the job. Because Tom has no workers' compensation insurance, Andy can file a workers' compensation claim against Diamond—even though he is not Diamond's employee. Because house painting is clearly a part of Diamond's regular business of constructing new housing, and Tom failed to get coverage, Andy is considered an employee of Diamond for purposes of workers' compensation insurance.

Because of these rules, it's very important to require any IC you hire to provide his or her own employees with workers' compensation coverage. Ask to see an insurance certificate proving that the employees are covered before you hire the IC. In many states, you can also call the workers' compensation agency to find out whether an IC has coverage for his or her employees.

If your own workers' compensation insurer audits your company and finds out that you have hired an IC who does not have coverage for his or her own employees, it is likely to classify those employees as your own employees. You will have to cover them on your policy and pay additional premiums. If this happens to you, you are legally entitled to request reimbursement from the IC. But if the IC has no money or can't be located, this legal right won't do you much good.

> **CAUTION**
> **California construction contractors beware.** In California, a construction contractor who hires unlicensed subcontractors or construction workers is automatically considered to be their employer for purposes of workers' compensation coverage, even if the workers are ICs under the usual common law rules. (Cal. Lab. Code § 2750.5.) If you're not sure whether the work you need done requires a license, contact the California Contractors State License Board—it decides who must be licensed to perform services in the construction industry in California. If a license is required, ask to see one before hiring a construction worker or subcontractor. For more details, refer to www.edd.ca.gov/pdf_pub_ctr/de231g.pdf.

Obtaining Coverage

Workers' compensation insurance must be purchased as a separate policy from a workers comp insurer. It can't be purchased as part of a business owner's liability insurance policy. Each state has its own rules about where employers may buy workers comp insurance. In a few states, all employers must buy their workers comp insurance from a "state fund" that has a monopoly on issuing such insurance in the state. In other states, workers' comp coverage may be purchased from either the state fund or from private insurers.

However, less desirable employers—those with bad claims histories or with very small payrolls—often have difficulty obtaining affordable workers' comp insurance from private carriers. In the states that have state workers' comp insurance funds, such funds are typically the insurer of last resort for businesses that cannot find coverage from a private insurer. In other states, less desirable employers can often find coverage only through the state's "assigned risk pool," in which coverage is pooled among a number of private insurers who are required by law to provide such coverage. Such coverage is typically more expensive than for workers comp insurance obtained through the regular voluntary private market.

Some states allow an employer to self-insure—a process that typically requires the business to maintain a hefty cash reserve earmarked for workers' compensation claims. Usually, this isn't practical for small businesses.

If private insurance is an option in your state, discuss it with an insurance agent or broker who handles the basic insurance for your business. Often, you can save money on premiums by coordinating workers' compensation coverage with property damage and liability insurance. A good agent or broker will be able to explain the mechanics of a state fund if that's an option or is required.

Hiring Household Workers and Family Members

Household Workers 184

    Federal Payroll Tax Status 184

    Payroll Tax Rules 186

    Insurance for Injuries to Household Workers 191

    Federal Minimum Wage and Overtime Regulations 192

    Immigration Requirements 197

Family Members as Workers 198

    Federal Payroll Taxes 198

    State Payroll Taxes 201

This chapter provides guidance if you have hired, or intend to hire, a person to work in or around your home—or if you hire a parent, spouse, or child to work at any location. It explains a host of rules that often make life a little easier for people who hire these types of workers.

Household Workers

Household workers include housecleaners, cooks, chauffeurs, housekeepers, nannies, babysitters, gardeners, private nurses, health aides, caretakers, and others who work in the home.

Federal Payroll Tax Status

You don't have to pay or withhold any federal payroll taxes (Social Security tax (FICA), federal unemployment taxes (FUTA), or federal income tax withholding (FITW)) for household workers who are ICs. However, you must pay FICA and FUTA for employee household workers whose salaries exceed certain amounts. (See below.)

You must apply the IRS test discussed in Chapter 3 to determine whether a household worker is an IC or employee for federal payroll purposes. Under this test, a worker is an employee if you have the right to control how the worker performs services. It's highly likely that any full-time household worker would be viewed as an employee under this test—you'll have a very hard time convincing the IRS that you don't have the right to control someone who works full-time in your home.

However, it is possible for part-time household workers to qualify as ICs under the test. The key is that the worker must be running an independent business.

> EXAMPLE: Anne hires David to clean her house every two weeks, a task that takes him about two hours. David has 20 clients in addition to Anne, provides his own cleaning equipment, and does the cleaning work when Anne is not home. Anne has the right to accept or reject the results David achieves in cleaning her home, but she does not supervise how he does the work. David is likely an IC under the IRS test.

Part-time housecleaners, cooks, chauffeurs, housekeepers, gardeners, caretakers, or maids may qualify as ICs under the common law test, but it seems likely that in-home child care workers would not, even if they work only part time. Even the most callous parent would probably insist on having the right to control how a babysitter, nanny, or similar worker cares for his or her children. Undoubtedly, most parents actually exercise this right. The only exception might be if you obtain the worker through an agency (see below). In addition, child care workers who provide care outside your home can be ICs.

The IRS Safe Harbor

Even if a household worker is an employee under the common law test, you won't have to pay employment taxes if you qualify for safe harbor protection. To qualify, you must satisfy two requirements:

- You must never have treated any worker performing similar services since 1977 as an employee for federal tax purposes. For example, if your household worker is a nanny, you must have never done any of the following for that worker or any other nanny you've hired since 1977: (1) withheld federal income tax or FICA tax from the worker's wages; (2) filed a federal employment tax return for the worker on IRS Form 942; or (3) filed a W-2 *Wage and Tax Statement* for the worker, whether or not tax was actually withheld.

- You must have had a reasonable basis for treating the worker as an IC. The most likely basis would be that the classification is a recognized practice of a significant segment of the worker's industry. After all, treating household workers as ICs has long been a common practice. However, as a practical matter, it can be hard to convince the IRS or courts that you have a reasonable basis for your IC classification. This is especially true today because eased tax reporting requirements that went into effect in 1995 have encouraged more people to treat household workers as employees.

(See Chapter 3 for a detailed discussion of the safe harbor.)

1099s Not Required for Household Workers

You don't need to file 1099-MISC forms reporting to the IRS your payments to household workers, no matter how much you paid them. You must file 1099s only when you hire a worker to work in your business. They need not be filed when you hire someone to work in your home for nonbusiness purposes—for example, to clean.

Workers Obtained Through Agencies

Generally, household workers obtained through an agency are not your employees if the agency is responsible for who does the work and how it is done. A babysitter you hire through a placement agency to come to your home to care for your child is not your employee if the agency sets and collects the fee, pays the sitter, and controls the terms of work—for example, provides the sitter with rules of conduct and requires regular performance reports. The agency is the sitter's employer, not you.

Independent Contractor Agreements

It's wise to have IC household workers sign independent contractor agreements before they start work. The agreement should make clear that the worker is an IC and you have no right to control the means and manner in which the work is performed—only the final results. (See Chapter 12 for guidance on creating your own IC agreements.)

Unfortunately, because many household worker relationships are informal, it may be difficult to get the worker to sign such an agreement. Do the best you can. It may be helpful to point out that such an agreement benefits the worker as well as you because it helps prevent possible disputes by setting forth the worker's duties and your payment obligations.

Payroll Tax Rules

Even if a household worker qualifies as an employee, you still may not have to pay federal payroll taxes.

Under the Social Security Domestic Employment Reform Act, you must pay FICA taxes for a household worker who qualifies as an employee under the common law test only if the worker is paid more than an annual threshold amount. The amount is adjusted annually to account for inflation. For 2014, it was $1,900. To find out the amount for subsequent years, refer to IRS Publication 926, *Household Employer's Tax Guide*. (You can get a copy from the IRS website at www.irs.gov.)

Employers who pay a household worker less than the threshold amount need not file federal tax forms for that worker. But household employees who earn less than the threshold still must pay their own income, FICA, and FUTA taxes unless their overall income is so low that they're not required to file a tax return.

If you do pay a household employee more than a certain amount per year, however, you must comply with a number of federal tax requirements. The following IRS chart summarizes these rules.

Federal Tax Requirements for Household Workers		
Type of Tax	**ICs**	**Employees**
FICA	None due	FICA tax is due if you pay cash wages of $1,900 or more during the year. But don't count wages you pay to: • your spouse • your child under age 21 • your parent (but see below for exception), or • any employee under age 18 (but see below for exception).
FUTA	None due	FUTA tax is due if you pay cash wages of $1,000 or more in any calendar quarter. But don't count wages you pay to: • your spouse • your child under age 21, or • your parent.
FITW	None due	FITW need not be withheld unless the employee requests it and you agree.

FICA Taxes

If you pay a household employee who is over 18 years of age $1,900 or more in cash wages in any year, you must withhold FICA taxes from the employee's earnings and make a matching contribution. Currently, the employer and employee must each pay an amount equal to 7.65% of the employee's wages.

You don't have to pay FICA taxes for household employees who were under 18 any time during the calendar year and for whom household service is not a primary occupation. But you must pay FICA if domestic service is the teenager's principal employment. In other words, FICA taxes are not due if a teenager works occasionally to earn extra money and not to earn a living. FICA must be paid, however, if a teenager does household work to earn a living. If a teenager is a student, providing household services is not considered to be his or her principal occupation.

> EXAMPLE 1: The Bartons hire Eve, a 17-year-old high school student, to babysit their children two or three times a month. The Bartons need not pay FICA for Eve even if they pay her $1,900 or more during the year.

> EXAMPLE 2: The Smiths hire Jane to provide child care services in their home. Jane is a 17-year-old single mother who left school and works as a child care giver to support her family. This is clearly her principal occupation. The Smiths must pay FICA for Jane if they pay her $1,900 or more during the year.

There is also a special exemption for family members. (See below.)

FUTA Taxes

If you pay a household employee $1,000 or more in cash wages during any calendar quarter—that is, any three-month period—you must also pay FUTA taxes. The rate varies from state to state depending on the amount of state unemployment taxes, but it is usually 0.6% of the first $7,000 of annual wages paid to an employee, or $42 per year.

⚠ **CAUTION**

Beware of changing state laws. Most states have already amended—or are expected to amend—their unemployment taxation laws to parallel the federal rule. Check with your state labor department to find out the current requirements in your state.

FITW Taxes

You don't have to withhold federal income taxes from a household employee's wages unless the employee requests it and the employer agrees. The same rule is followed under most state income tax laws.

It's unlikely that a worker would make such a request because most workers prefer not to have tax withheld from their paychecks. But if a worker does ask you to withhold income tax, it's probably not in your interest to agree; it will only create extra bookkeeping headaches for you.

Paying FICA and FUTA Taxes

You are responsible for withholding your household employee's share of FICA taxes and paying your own. Withholding means you deduct the taxes due from the worker's pay and keep it in your bank account. IRS Publication 926, *Household Employer's Tax Guide,* contains a table showing you how much you should withhold from the wages you pay a household employee. Instead of withholding the employee's share of these taxes, you can pay them from your own funds. Obviously, if you do this, you should reduce the employee's compensation to make up for the tax payments you're making on the employee's behalf.

Federal unemployment or FUTA taxes work differently. You cannot withhold FUTA taxes from an employee's wages. You must pay them yourself.

When you file your federal income tax return, you must include as income all the FICA and FUTA taxes due on the wages you paid your household employee. The amount you owe on this additional "income" is due to the IRS with your tax return by April 15. You report your household employment taxes on IRS Schedule H, which you attach to your Form 1040.

If you have several household employees or pay them a lot, you could have substantial extra taxes due when you file your tax return. You can avoid this by paying estimated taxes during the year to the IRS to cover the amount of employment taxes due. Alternatively, if you are employed, you can have your employer increase the amount of federal tax it withholds from your paychecks.

If you don't pay estimated tax or have enough tax withheld from your paychecks, you may have to pay the IRS an estimated tax penalty. Generally, you'll have to pay a penalty if the amount you have withheld from your paychecks or pay as estimated tax during the year is less than 90% of your total tax due for the current year.

RESOURCE

Need more information on withholding? See IRS Publication 505, *Tax Withholding and Estimated Tax*, which you can get at www.irs.gov.

If you own a business as a sole proprietor or if your home is on a farm and you have business or farm employees, you can choose between two ways of paying your household employment taxes. You can pay them with your federal income tax as described above, or you can include them with your federal employment tax deposits or other payments for your business or farm employees. You may not deduct wages and employment taxes paid to your household employees on your Schedule C or F; those deductions apply only to wages and taxes paid for business and farm employees.

If you withhold or pay FICA taxes or withhold federal income tax, you must file IRS Form W-2 after the end of the year. To complete Form W-2, you will need both an employer identification number and your employee's Social Security number.

CAUTION

Household employers must obtain federal ID numbers. If you hire a household employee, the IRS requires you to obtain a federal employer identification number or EIN. An EIN is a nine-digit number the IRS assigns to employers for tax filing and reporting purposes. EINs are free and easy to obtain. Use your EIN on all your employment tax returns, employment tax checks, and other employer-related documents you send the IRS.

The fastest and easiest way to obtain an EIN is electronically. Go to the IRS website at www.irs.gov and enter keyword "EIN." You'll find a link to take you to the online EIN application page. You don't need to fill out IRS Form SS-4, *Application for Employer Identification Number.* Instead, you answer a series of questions. After the IRS validates the information, it will issue you an EIN immediately. You can then download, save, and print your EIN confirmation notice.

If you don't want to apply online, you can obtain your EIN by calling the IRS at 800-829-4933. Or you can mail a completed Form SS-4 to the appropriate IRS service center listed in the form's instructions. The IRS will mail the EIN to you in about four weeks.

Insurance for Injuries to Household Workers

Household workers can become injured on the job. For example, a baby-sitter could slip on a toy carelessly discarded by your child and suffer a back injury. The sitter will undoubtedly look to you to pay the medical and other expenses caused by the job-related accident.

You could be liable for such injuries. For example, you'll generally be liable for a work-related injury to a household worker that is caused by your or your family members' negligence, or unsafe conditions in your home. But even if you're not liable, you could still be sued by an injured household worker and have to hire an attorney and pay other legal expenses.

Paying for these expenses out of your own pocket could prove ruinous. You should have insurance to cover them. This normally takes the form of a homeowner's insurance policy that also provides workers' compensation coverage for injuries to household employees.

Coverage for Household ICs

If you own your home, you probably already have a homeowner's insurance policy (all lenders require them). Homeowner's policies contain liability coverage that insures you if a household worker who is an IC sues you for bodily injury or property damage occurring at your home. Your insurance company will pay the costs of defending such a lawsuit and pay any damages up to the policy limits. It will also pay the injured person's medical expenses.

Coverage for Household Employees

Injuries to household employees may not be covered by your homeowner's policy. Such policies typically exclude coverage for injuries to employees. Instead, you have to purchase workers' compensation insurance. (See Chapter 6 for information on workers' compensation requirements.)

If you only use ICs and never hire a household employee, you don't need workers' compensation insurance. Unfortunately, it can be difficult to know for sure whether a worker is an employee or IC for workers' compensation purposes. States use different tests to classify workers for this purpose. These rules can be complex and difficult to apply. (See Chapter 6.) Don't gamble that a household worker is an IC. Your homeowner's insurer may disagree with you and claim that the worker is your employee. It will then deny coverage to an injured worker under the bodily injury and medical payment provisions of your policy.

You can usually obtain workers' compensation coverage for household employees from your homeowner's insurer. Your homeowner's policy may already include this coverage. Or you may have to specifically ask for it and pay extra. Check your policy or ask your insurance agent about it. If your policy doesn't already include this coverage, you'll need to purchase a rider or an endorsement covering household employees.

Renters

If you're a renter, you should obtain a renter's policy with this same coverage. Don't assume that your landlord's insurance will cover your household workers.

Federal Minimum Wage and Overtime Regulations

The federal Fair Labor Standards Act (FLSA) gives most types of employees the right to be paid at least the federal minimum wage and time-and-a-half for overtime. The FLSA is enforced by the Department of Labor, which may impose fines against employers who violate it.

You may be surprised to discover that the FLSA applies to household employees if they:

- receive at least $1,900 in cash wages in a calendar year from one employer, or
- work a total of more than eight hours a week for one or more employers.

For excellent guidance on these issues, see the U.S. Department of Labor website at www.dol.gov/whd/homecare/index.htm.

Classifying Workers Under the FLSA

The Department of Labor uses an economic reality test to determine a worker's status for FLSA purposes. (See Chapter 8 for a discussion of this test.) To qualify as an IC, a household worker will generally have to provide services for several different households simultaneously and be able to show some opportunity for profit or loss. Profit or loss can be shown where a worker earns a set fee instead of an hourly wage, or where a worker has business expenses that could exceed business income—for example, salaries for assistants or equipment costs.

Of course, you cannot closely supervise the work of a household worker and expect him or her to qualify as an IC. Instead, your control must be limited to accepting or rejecting the worker's final results. For example, you can tell a gardener to mow your lawn and rake leaves, but you can't supervise how he or she does the work. There are some types of household workers, such as child care providers, whom you will want to supervise closely—which means they are not ICs.

Any worker who works solely for you and makes no attempt to obtain other clients or customers will almost surely be viewed as your employee by the Department of Labor. For example, a live-in housekeeper or child care provider is almost certainly an employee.

These federal wage and overtime regulations apply only to household employees, not ICs. Unfortunately, it can be hard to know for sure whether a household worker is an employee or IC under the FLSA. If you're not sure whether a household worker is an employee or IC, the safest course is to assume he or she is an employee and obey the minimum wage and overtime rules. (See below.) These rules do not place a great financial burden on you.

Minimum Wage Laws

The federal minimum wage is $7.25 per hour. If your state has established a higher minimum wage, you must pay that amount. In California, for example, the minimum wage is $9.00 per hour (and is scheduled to go up to $10 per hour on January 1, 2016). Nineteen other states also have minimum wage rates higher than the federal minimum. Some states allow employers to pay inexperienced employees a lower training wage for the first few months they are on the job. For example, California employees may be paid 85% of the current state minimum wage during their first 160 hours of employment in occupations in which they have no previous similar or related experience.

You can find your state's minimum wage by checking the U.S. Department of Labor website at www.dol.gov (click "Wages," "Minimum Wage," then "Minimum Wage Laws in the States") or checking with your state labor department. In the states that have a minimum wage lower than the federal rate, the federal rate controls.

You must pay the minimum wage to any nonexempt domestic employee who:

- earns at least $1,900 in wages from one employer in any calendar year, or
- works more than eight hours a week for one or more employers.

If a household employee works less than eight hours a week for you, but works for others as well, you'll have to find out how many hours he or she works for other employers.

You must pay a household employee the minimum wage regardless of whether you pay by the hour or with a regular salary. You must pay minimum wage for each hour worked, including all hours an employee must be on duty at your home or at any other prescribed workplace, such as a vacation house. However, you can take a credit for the value of room and board you provide to a household employee.

How Room and Board Figure Into Minimum Wage

Under the FLSA, employers may take a credit against minimum wage requirements for the reasonable cost or fair value of food and lodging or other facilities customarily furnished to an employee. But an employer may take this credit only when the employee voluntarily agrees to the arrangement.

Federal and state regulations define appropriate meal and lodging credits. In California, for example, when credit for lodging is used to meet part of the employer's minimum wage obligation, no more than $37.63 per week may be credited for a room occupied by one person.

Compensation for Overtime

Most household employees must be paid overtime at the rate of one and one-half times their regular wage rate for all hours worked beyond a 40-hour workweek. In computing overtime pay, you must treat each workweek separately—that is, you can't average hours over two or more weeks.

Some states, including California, have a daily overtime standard. If you live in one of these states, you must pay overtime if your employee works more than eight hours in one day.

Household Workers Exempt From Minimum Wage and/or Overtime Requirements

Some household workers are exempt from the FSLA overtime and/or minimum wage requirements, even though they are classified as employees. However, under tighter rules that go into effect on January 1, 2015, these exemptions may be claimed only by the individual, family, or household using the services, as opposed to a third-party employer, such as a home health care agency. The regulations also impose recordkeeping requirements on employers of these workers (other than where casual babysitters are at issue).

Exempt household employees include the following:

Casual babysitters. Casual babysitters are exempt from federal minimum wage and overtime requirements. A "casual babysitter" is a person who doesn't depend on his or her income from babysitting for a livelihood. Thus, this exemption doesn't apply to trained personnel whose vocation is babysitting or caring for others—for example, registered and practical nurses.

As a rule, casual babysitters may regularly work no more than 20 hours per week, though they may work more hours on an "irregular or intermittent" basis. Casual babysitters must spend the bulk of their time caring for children—they must spend less than 20% of their total hours on general household work.

The casual babysitter exemption also applies to a sitter who accompanies one's family on vacation, provided that the vacation doesn't exceed six weeks.

Live-in domestic employees. Domestic service workers who reside in the employer's home and are employed by an individual, family, or household are exempt from the overtime pay requirement, but they must be paid at least the federal minimum wage for all hours worked. Thus, you don't have to pay overtime to an au pair, a housekeeper, or another household employee who lives in your home. You must, however, pay the minimum wage.

This exemption applies only to workers who are employed directly by the family they live with, not to those employed by third-party agencies.

Companionship services employees. Certain household employees providing companionship services are exempt from the FSLA overtime and minimum wage requirements. To be exempt, the worker must be employed directly by the family (not by a third-party agency, for example) and have duties mostly limited to "fellowship and protection," such as spending time together watching television, engaging in hobbies, or taking walks. Under new regulations that go into effect on January 1, 2015, the companion may spend no more than 20% of his or her time performing "care"—that is, helping with activities of daily living such as dressing, feeding, bathing, meal preparation, driving, and housework. And, someone who performs medical tasks or does chores that benefit other members of the household (such as laundry, preparing meals, or cleaning) isn't covered by the exemption.

Immigration Requirements

Many household workers in the United States are immigrants. Some work illegally —that is, they are not U.S. citizens and don't have a green card or other documentation of their legal status. The federal government is cracking down on people who illegally hire undocumented household employees.

All employers, including those who hire most types of household employees, are required to verify that the employee is either a U.S. citizen or national, or a legal alien authorized to work in the United States.

You are not required to verify citizenship when you hire an IC. The government uses the common law right-of-control test to determine whether a worker is an employee or IC for immigration purposes. (See Chapter 2 for a discussion of the common law test.) A worker who qualifies as an IC for tax purposes will likely be an IC for immigration purposes as well.

Verification is also not required for employees who work in your home on only a sporadic basis—for example, a babysitter you hire now and then to sit for a few hours. Nor is verification necessary for employees of domestic agencies—for example, a housekeeper you hire from an agency where the worker is the agency's employee and you pay the agency, not the worker directly.

Although you are not required to verify the immigration status of ICs or others coming within these exceptions, it is still illegal for you to hire any worker whom you know to be an illegal alien.

RESOURCE

Want to help a household worker become legal? See *U.S. Immigration Made Easy*, by Ilona Bray (Nolo). Also contact the United States Citizenship and Immigration Services department (USCIS) about any federal programs or special visas that apply to nannies or au pairs. The USCIS (formerly the Immigration and Naturalization Service, or INS) has an informative website at www.uscis.gov.

Family Members as Workers

Family members work with and for each other all the time. If a family member is an IC under the tests discussed in this book, then that's great—you don't have to worry about payroll taxes and unemployment compensation (depending on the test, of course). And even if the family member is your employee under these tests, you may still escape payroll taxes under special state rules.

Federal Payroll Taxes

To determine whether a family member is an IC for IRS purposes, follow the steps described in Chapter 3. If the family member is an IC, then you can stop there. If the family member is an employee, however, you may still escape federal payroll taxes (FICA and FUTA). Read on.

Children Employed by Parents

A parent need not pay FUTA taxes for services performed by a child who is younger than 21, regardless of the type of work the child does.

FICA taxes need not be paid for a child younger than 18 who works for a parent in a trade or business, or a partnership in which each partner is a parent of the child. If the services are for work other than a trade or business—such as domestic work in the parents' home—the parent does not have to pay FICA taxes until the child reaches 21.

> **EXAMPLE:** Lisa, age 16, works in a bakery owned by her mother and operated as a sole proprietorship. Although Lisa is an employee under the IRS test, her mother need not pay FUTA for Lisa until she reaches 21 and need not pay FICA taxes for her until she reaches 18.

However, these rules do not apply—and FICA and FUTA must be paid—if a child works for:

- a corporation, even if it is controlled by the child's parent
- a partnership, even if the child's parent is a partner, unless each partner is a parent of the child, or
- an estate, even if it is the estate of a deceased parent.

EXAMPLE: Ron works in a bicycle repair shop that is half owned by his mother and half owned by her partner, Ralph, who is no relation to the family. FICA and FUTA taxes must be paid for Ron because he is working for a partnership and not all the partners are his parents.

If a child is paid regular cash wages as an employee in a parent's trade or business, he or she may be subject to federal income tax withholding regardless of age.

One Spouse Employed by Another

If one spouse pays another wages to work in a trade or business, the payments are subject to FICA taxes and federal income tax withholding, but not to FUTA taxes.

EXAMPLE: Kay's husband, Simon, is a lawyer with his own practice. Kay works as his secretary and is paid $1,500 per month. Simon must pay the employer's share of FICA taxes for Kay and withhold employee FICA and federal income taxes from her pay.

However, neither FICA nor FUTA need be paid if the spouse performs services other than in a trade or business—for example, domestic service in the home.

EXAMPLE: Jill is a medical doctor with a busy practice. Her husband, Bob, stays home and takes care of the house and children. Jill gives Bob $1,000 a month as walking around money. These payments are not subject to any federal payroll taxes—FICA, FUTA, or FITW.

But these rules do not apply—and FICA, FUTA, and FITW must all be paid—if a spouse works for:
- a corporation, even if it is controlled by his or her spouse
- a partnership, even if his or her spouse is a partner, or
- an estate, even if it is the estate of a deceased spouse.

EXAMPLE: Laura's husband, Rob, works as a draftsperson in Laura's architectural firm. The firm is set up as a corporation solely owned and controlled by Laura. The corporation must pay FICA, FUTA, and FITW for Rob.

Parent Employed by Child

The wages of a parent employed by a son or daughter in a trade or business are subject to income tax withholding and FICA taxes.

EXAMPLE: Don owns and operates a restaurant and employs Art, his father, as a part-time waiter. Because the restaurant is a business, Don must pay the employer's share of FICA taxes for Art and withhold employee FICA and federal income taxes from his pay.

FICA taxes do not have to be paid if the parent's services are not for a trade or business—for example, domestic services in the home. However, this rule is subject to one exception. Wages for domestic services by a parent for a child are subject to FICA taxes if:

- the parent cares for a grandchild (that is, the parent's child's child) who is either younger than 18 or requires adult supervision for at least four continuous weeks during a calendar quarter due to a mental or physical condition, and
- the parent's child is a widow or widower, divorced, or married to a person who, because of a physical or mental condition, cannot care for the grandchild.

EXAMPLE: Sally is a divorcee with two small children who live with her. Sally works during the day so she hires Martha, her mother, to care for the children during working hours. Sally pays Martha $250 a week. Sally must pay the employer's share of FICA taxes for Martha and withhold employee FICA and federal income taxes from her pay.

You do not have to pay FUTA taxes when you hire your parent to perform household services.

State Payroll Taxes

State payroll taxes consist of unemployment compensation, which employers are required to pay directly to a state fund, and state income tax, which employers must withhold from employees' paychecks and remit to the state taxing authority. However, you will not have to pay taxes on services performed by a spouse or minor child or, with few exceptions, services performed by a parent.

Unemployment Compensation

Every state exempts from unemployment compensation coverage services performed by an employer's minor child or spouse.

In over half the states, a minor child is one under 21 years old. In most of the other states, a minor is a child under 18. In Wyoming, the age is 19.

State Income Taxation

All states except Alaska, Florida, Nevada, New Hampshire, South Dakota, Tennessee, Texas, Washington, and Wyoming tax earned income. If a family member is an employee of your business and is paid regular wages, you may have to withhold state income taxes from his or her pay. Check with your state's tax authority.

No income tax withholding is required for family members who qualify as ICs under your state's income tax law.

CHAPTER

8

Health, Safety, Labor, and Antidiscrimination Laws

Obamacare (Affordable Care Act) 206

    "Large Employers" 207

    Calculating the Number of Employees 207

    ICs Not Covered by Mandate 209

    The Common Law Test for Employment Status 211

    Penalties Under the ACA 211

    No-Coverage Penalty for 2015 212

    No-Coverage Penalty for 2016 and Beyond 213

    Avoiding the No-Coverage Penalty 214

Federal Wage and Hour Laws 215

    When the FLSA May Apply 216

    Covered Businesses 217

    Workers Exempt From Overtime and Minimum Wage Requirements 217

    Classifying Workers Under the FLSA 219

    Avoiding Problems 222

    Record-Keeping Requirements 222

Federal Labor Relations Laws 222

    ICs Are Not Covered 223

    Employees Exempt From the NLRA 223

    Determining Worker Status 223

Family and Medical Leave Act 223

Fair Credit Reporting Act 224

Antidiscrimination Laws 225

    Federal Antidiscrimination Laws 225

    State Antidiscrimination Laws 228

Worker Safety Laws..229

  OSHA Coverage of ICs..229

  Importance of Maintaining a Safe Workplace..230

Immigration Laws 230

E mployees enjoy a wide array of rights under federal and state health, safety, labor, and antidiscrimination laws. Among other things, these laws:

- impose minimum wage and overtime pay requirements on employers
- make it illegal for employers to discriminate against employees on the basis of race, color, religion, gender, age, disability, or national origin
- protect employees who wish to unionize, and
- make it unlawful for employers to knowingly hire illegal aliens.

In addition, the Obamacare "employer mandate," requiring large employers to provide health insurance to their full-time employees or face substantial penalties, takes effect in 2015.

In recent years, a growing number of employees have brought lawsuits against employers alleging violations of these laws. Some employers have had to pay hefty damages to their employees. In addition, various watchdog agencies, such as the U.S. Department of Labor and the U.S. Equal Employment Opportunity Commission, have authority to take administrative or court action against employers who violate these laws.

One of the advantages of hiring ICs is that few of these laws apply to them.

RESOURCE

For more information about health, safety, labor, and antidiscrimination laws as they apply to employees, refer to *The Manager's Legal Handbook.* If you want detailed information about these laws, including the text of the laws and government resources, refer to *The Essential Guide to Federal Employment Laws.* Both are by Amy DelPo and Lisa Guerin, and both are published by Nolo.

Obamacare (Affordable Care Act)

The Affordable Care Act (ACA), more commonly referred to as Obamacare, is sweeping federal legislation designed to ensure that all Americans have access to affordable health insurance coverage. Among its rules, Obamacare requires that most Americans have at least "minimum essential health care coverage" for themselves and their dependents—otherwise, they must pay a tax penalty. Obamacare also establishes health insurance exchanges through which individuals and small businesses can purchase health insurance coverage and, depending on their income, obtain tax subsidies to help pay for such coverage.

The IRS administers the tax provisions in Obamacare, including imposition of the tax penalties described below. The U.S. Department of Health and Human Services (HHS) is in charge of the health exchanges and other nontax portions of the law.

Most of Obamacare's provisions took effect on January 1, 2014. However, one important portion—the employer mandate—doesn't begin to take effect until January 1, 2015. Starting on that date, the mandate will be phased in over two years. In 2015, large employers will be required to provide 70% of their full-time employees and those employees' families (a qualification regarding employee families is discussed below) with minimum essential health care insurance. (Such insurance must pay for at least 60% of covered services.) Employers can require that employees contribute toward their insurance coverage, but they can't require them to pay more than 9.5% of their household income toward it. (The percentage of employees for whom employers must provide insurance increases in 2016.)

Large employers who fail to comply with the coverage mandate will have to pay a penalty to the IRS. Small employers—that is, employers who don't qualify as large employers—aren't subject to the employer mandate, and need not provide any health insurance coverage to their employees.

As a practical matter, few businesses will be affected by the employer mandate. Only slightly over 3% of all businesses in the United States—approximately 200,000 out of six million—will be subject to the mandate when it takes full effect in 2016. And 96% of these already provide their full-time employees with health insurance.

"Large Employers"

In terms of Obamacare, the crucial question for a hiring firm is whether it is a "large employer." The definition for 2015 differs from that for 2016 and later. The changing definition gives medium-sized business an additional year to get ready for the mandate.

Rule for 2015. For 2015 only, a large employer is any organization that employed, on average, a combination of 100 or more full-time and "full-time equivalent" (FTE) employees during six months or more of 2014. Employers with 50 to 99 employees during 2014 are exempt from the large-employer mandate for 2015 unless they:

- reduce their workforce below 100 solely to avoid providing health coverage, or
- eliminate or materially reduce the health coverage they offered their employees as of February 9, 2014.

Employers with 50 to 99 employees must file with the federal government a certification establishing that they haven't taken either of the foregoing actions.

Rule for 2016 and later. For 2016 and later, a large employer is any organization that employed, on average, a combination of 50 or more full-time and FTE employees the preceding year. Thus, any employer with 50 or more employees in 2015 will be a large employer subject to the mandate in 2016.

Calculating the Number of Employees

Whether a hiring firm is a large employer for any given year is determined by counting the number of employees it had in the preceding year. Thus, hiring firms need to keep track of the number of full-time and FTE employees they have during 2014 to determine whether they will be deemed large employers on January 1, 2015, when the Obamacare employer mandate kicks in.

Ordinarily, you look at the entire calendar year to determine how many full-time and FTE employees an employer had on average. Thus, you add the total number full-time and FTE employees for each month in the preceding calendar year and divide by 12. However, for 2015 only, employers have the option of looking at any consecutive period of at least six months during 2014, rather than the whole year. This rule is intended to make compliance easier for the first year the employer mandate takes effect. For example, an employer can use a period of at least six months through October 2014 to determine whether it meets large-employer status; if it does, it can use November and December 2014 to make any needed adjustments to its plan (or to establish a plan).

Full-time and FTE employees must be counted. Both full-time and FTE employees count toward the employee threshold.

- "Full-time" denotes an employee who works on average at least 30 hours per week in a given month. (For these purposes, 130 hours of service in a calendar month is the equivalent of 30 hours per week).
- The number of FTE employees is determined by accounting for all the hours worked by part-time employees. To make this determination for any month, you divide the total number of hours worked by employees who aren't full-time by 120. For example, two employees, each of whom works 15 hours per week (60 hours per month), are the equivalent of one full-time employee $((2 \times 15 \times 4) \div 120 = 1)$.

EXAMPLE 1: During 2014, Acme, Ltd. has 80 full-time employees working 40 hours per week, along with 40 part-time employees working 15 hours per week. The 40 part-time employees count as 20 FTE employees $((40 \times 15 \times 4) \div 120 = 20)$. Thus, Acme has a combined 100 full-time and FTE employees and is a large employer for 2015.

EXAMPLE 2: During 2015, ABC, Inc., has 40 full-time employees working 40 hours per week, along with 20 part-time employees working 15 hours per week. The 20 part-time employees count as 10 FTE employees $((20 \times 15 \times 4) \div 120 = 10)$. ABC has a combined 50 full-time and FTE employees and is a large employer for 2016.

Seasonal employees. Employers don't have to count seasonal employees who work six months or less during the year toward the large-employer threshold. (But, as discussed above, the hours of nonseasonal, part-time employees count toward the threshold calculation.) Even though seasonal employees don't count toward the threshold, firms that qualify as large employers must offer health care coverage to all full-time employees, including seasonal employees who work full time in any month.

Combining employees. Companies that have a common owner or are otherwise related generally are combined and treated as a single employer for Obamacare purposes. Thus, they would be combined to determine whether they collectively employ at least 100 or 50 full-time and/or FTE employees. If the combined total meets the applicable threshold, then each separate company is subject to the employer mandate, even those companies that individually do not employ enough employees to meet the threshold. For example, if an individual owns 80% or more of two businesses that are separate legal entities, the total number of full-time and FTE employees of that employer is based on the full-time and FTE employees in both businesses combined.

As a result, an employer can't avoid being classified as a large employer by dividing up its operations into a number of separate legal entities under common ownership and then counting each as a separate employer for Obamacare purposes.

ICs Not Covered by Mandate

Independent contractors aren't included in Obamacare's employer mandate— in other words, large employers need not provide health insurance coverage to any ICs they hire. Moreover, ICs' hours worked don't count toward the 50- or 100-employee threshold for determining whether a hiring firm is subject to the employer mandate.

However, ICs themselves are subject to Obamacare's requirement that people have minimally adequate health insurance. Unless exempt, ICs must obtain such insurance for themselves and their dependents or face a tax penalty. ICs can purchase health insurance through the state health insurance exchanges established under Obamacare; those whose family incomes are less than 400% of the federal poverty rate can obtain subsidies to help purchase health insurance. Hiring firms should direct ICs to the Obamacare website at www.healthcare.gov to obtain more information about the program. A hiring firm should not help or advise an IC regarding health insurance coverage. This is something that ICs—as independent businesspeople—are supposed to handle themselves.

Determining Employee Hours

2014 is the time for all hiring firms to figure out whether they are large employers for Obamacare purposes—that's how they will know whether they are subject to the employer mandate starting on January 1, 2015.

Employers must keep track of all the hours worked by their full- and part-time employees during each month of the year. They must track the actual hours worked by employees paid by the hour. For salaried and other nonhourly employees, an employer may choose to track actual hours worked, or use a days- or weeks-worked "equivalency method." Under this method, the employee is automatically deemed to work eight hours per day or 40 hours per week.

For employers with stable workforces, it might be a simple task to total the hours worked by all employees each month. However, for employers with seasonal employees (who don't count) and employees with variable hour-schedules, this task can be very complicated. To make things easier for employers, Obamacare includes an optional "look-back" measurement method in which you average variable-hour or seasonal employees' hours over a period of three to 12 months.

Of course, the IRS may determine that a worker classified as an IC by a large employer is actually an employee. In that case, the employer will have to establish that worker's hours. This calculation should be fairly straightforward for workers who were paid on an hourly basis. But, it could be tougher where the workers were paid on a per-project, commission, or other nonhourly basis. Presumably, worker testimony and available records would provide the basis for reconstructing the misclassified worker's hours.

The Common Law Test for Employment Status

IRS regulations provide that the common law test for employment status will be used to determine whether workers qualify as employees for Obamacare purposes. (79 Fed. Reg. 8544-01.)

This is the familiar test employed by the IRS and many other government agencies, discussed in detail in Chapter 2. The common law test hinges on whether the employer has the right to control the worker. The right of control is determined by evaluating numerous factors, none of which is individually conclusive.

Several categories of workers aren't classified as employees, either pursuant to the common law test or because of statutory-IC designation:

- leased employees (see Chapter 10)
- direct sellers who qualify as statutory ICs (see Chapter 3)
- licensed real estate agents who qualify as statutory ICs (see Chapter 3)
- sole proprietors
- partners in partnerships
- 2%-or-more shareholders in S corporations, and
- bona fide volunteers for government entities or tax-exempt organizations—for example, volunteer firefighters and other volunteer emergency responders.

Penalties Under the ACA

Large employers who fail to comply with the Obamacare employer mandate—that is, who don't provide adequate coverage to the required percentage of their employees and those employees' dependents—are subject to a significant no-coverage penalty, officially called the "employer shared responsibility payment." As discussed below, the rules for this penalty differ for 2015 and for 2016 and later. The no-coverage penalty is an excise tax and is not tax deductible, unlike contributions an employer makes for its employees' health insurance coverage.

Small employers whose full-time employee count is anywhere near the large employer threshold (100 or more employees in 2014 for purposes of 2015, 50 or more in 2015 for purposes of 2016) need to be very careful about classifying additional workers as ICs. If the IRS later determines that the workers should have been classified as ICs, the small employer could then become a large employer under Obamacare and be subject to the no-coverage penalty.

The Affordable Care Act also contains a $100-per-day-per-employee general noncompliance penalty. Employers who don't correct compliance failures within 30 days of discovery must self-report and pay the penalty on IRS Form 8928, *Return of Certain Excise Taxes Under Chapter 43 of the Internal Revenue Code.* Correction of the failure requires that the employer put the employee back in the same financial position he or she would have been in had the failure not occurred—presumably, the employee would have to be paid to make up for the failure to provide the mandated health insurance coverage.

For additional guidance on these complex penalties, review the IRS FAQs at www.irs.gov/uac/Newsroom/Questions-and-Answers-on-Employer-Shared-Responsibility-Provisions-Under-the-Affordable-Care-Act.

No-Coverage Penalty for 2015

For 2015 only, large employers (those with a combination of 100 or more full-time and FTE employees in 2014) will be subject to a no-coverage penalty only if they fail to offer minimally adequate health insurance to at least 70% of their full-time employees. No penalty is due for failure to offer coverage to part-time employees. In addition, for 2015 only, large employers need not offer coverage to their full-time employees' dependents so long as they are taking steps to arrange for such coverage to begin in 2016.

The penalty applies if at least one full-time employee receives a premium tax credit for enrolling in health insurance through a state health insurance exchange.

The amount of the penalty is $2,000 per year ($167 per month) times the total number of full-time employees minus 80 employees. (FTE employees don't count in this calculation.) If an employer offers coverage for some months but not others during the year, the payment is computed separately for each month for which coverage was not offered.

> **EXAMPLE:** During 2014, Acme, Inc. had 96 full-time employees and four full-time workers it classified as ICs. Since it claimed to have fewer than 100 employees in 2014, it concluded it was not a large employer for 2015 and was not subject to the employer mandate that year. It provided no health insurance to any of its employees during 2015. During 2015, ten of its employees obtained their own subsidized coverage through their state exchange. The IRS later audits Acme for 2014 and determines the four ICs should have been classified as employees. This means Acme had 100 full-time employees during 2014 and was a large employer in 2015 subject to the employer mandate. The IRS imposes a no-coverage penalty on Acme for 2015. Acme had 100 full-time employees during 2015, so the penalty is $40,000 ($2,000 x (100-80)).

No-Coverage Penalty for 2016 and Beyond

For 2016 and later, large employers (those with a combination of 50 or more full-time and FTE employees in the preceding year) must provide adequate coverage to at least 95% of their employees (including employees' dependents). If a large employer fails to do so, the no-coverage penalty is imposed, again, if least one full-time employee receives a premium tax credit for enrolling in health insurance through a state health insurance exchange.

The amount of the no-coverage penalty for 2016 and later is $2,000 per year ($167 per month) times the total number of full-time employees minus 30 full-time employees. (Again, FTE employees don't count in this calculation.) If an employer offers coverage for some months but not others during the year, the payment is computed separately for each month for which coverage was not offered.

> **EXAMPLE:** During 2015, ABC, Inc. had 48 full-time employees and two full-time workers it classified as ICs. Since it claimed to have fewer than 50 full-time and FTE employees in 2015, ABC concluded it was not a large employer for 2016. It provided

no health insurance to its employees during that year. Five of its employees obtained their own subsidized coverage for 2016 through their state exchange. The IRS later audits ABC and determines the two ICs should have been classified as employees during 2015. This means that ABC had 50 full-time employees during 2015 and was a large employer subject to the employer mandate in 2016. The IRS imposes an Obamacare no-coverage penalty on ABC for 2016. The company had 50 full-time employees during 2016, so the penalty is $40,000 ($2,000 x (50-30)).

Avoiding the No-Coverage Penalty

Because (1) no penalty is due for failure to provide coverage to part-time employees, and (2) a business's first 80 full-time employees don't count toward the 2015 no-coverage penalty, a business that doesn't provide health insurance coverage will not be subject to the penalty in 2015 if it had 80 or fewer full-time employees that year, regardless of how many FTE employees it had. For 2016 and later, the magic number of full-time employees goes down to 30.

> **EXAMPLE:** Acme, Inc. is large employer. In 2016 it has 30 full-time employees and 100 FTE employees. It provides none of them with health insurance. Its no-coverage penalty is $0. The penalty is equal to $2,000 times the number of full-time employees minus the first 30 full-time employees. Acme's 30 full-time employees minus 30 is zero, so no penalty is due.

Because of this loophole, some businesses have already reduced their number of full-time employees to 30 or fewer. Under Obamacare, a full-time employee must work an average of 30 or more hours per week (or 130 hours per month). So, some employers have cut their employees' hours to less than 30 per week so they won't be counted as full-time employees for purposes of Obamacare penalties for any future year.

Section 530 Provides No Relief From Obamacare Penalties

The Section 530 safe harbor rules provide hiring firms that qualify with relief from IRS penalties when they misclassify workers for purposes of employment taxes. (See Chapter 3 for a detailed discussion of Section 530.)

However, Section 530 applies only to employment taxes, not other types of taxes. The Obamacare penalties are legally classified as "excise taxes," so Section 530 doesn't apply to them. Industry groups requested that Section 530 be made applicable to Obamacare penalties, but the IRS expressly refused to do so in its final regulations.

Federal Wage and Hour Laws

The main federal law affecting workers' pay is the federal Fair Labor Standards Act, or FLSA (29 U.S.C. § 201 and following), which establishes a national minimum wage and overtime standards for covered employees.

Most businesses are covered by the FLSA, but not all workers are included in its coverage. ICs are not subject to this law. Nor are employees who fall within any of the several exempt categories discussed below.

CAUTION

Don't forget state laws. This discussion pertains only to the federal wage and hour law, not state laws. Most states have their own labor laws, and some give employees more rights than the FLSA. You must comply with whichever law—federal or state—gives your employees more protection. Find links to all 50 state labor departments and other state agencies involved with workers at www.dol.gov/dol/location.htm.

When the FLSA May Apply

You are required to pay employees working for you the federal minimum wage unless your state minimum is higher, in which case you must pay the latter. One possible exception is for household workers. (See Chapter 7 for guidance on hiring household workers.)

The FLSA also requires you to pay all nonexempt employees one and one-half times their regular rates of pay for all hours of work over 40 hours in a week. If you've classified as ICs workers who are really nonexempt employees under the FLSA, you will likely have to pay each misclassified worker this overtime premium for all hours worked in excess of 40 per week during the previous two or three years. This could be a substantial sum if your workers regularly put in long work weeks. If you refuse to pay, you could face legal action by the Labor Department or by the affected workers. You could also be held personally liable for FLSA violations.

The U.S. Department of Labor recently hired hundreds of investigators and other enforcement staff to undertake a "Misclassification Initiative" designed to recover back wages and other benefits from employers who misclassified their workers as ICs. (For details, visit www.dol.gov/whd/workers/misclassification/.) The Department of Labor also relies on complaints by disgruntled workers who believe they're entitled to overtime pay. Informants' identities are kept confidential, so workers really have nothing to lose if they think they might qualify as employees and be entitled to the protection of the FLSA.

Before trying to determine how workers will be classified by the Department of Labor, you should first see whether either your business or workers are exempt from FLSA coverage. You don't need to worry about the Department of Labor if your business or workers are exempt.

Covered Businesses

Your business is covered by the FLSA if you take in $500,000 or more in total annual sales or if you're engaged in interstate commerce. This covers nearly all workplaces, because the courts have broadly interpreted interstate commerce to include, for example, regularly using the U.S. mail to send or receive letters to and from other states or making or accepting telephone calls to and from other states.

If your business is covered by the FLSA and your employees are not exempt from the FLSA, they must be paid time-and-a-half for overtime. This is so whether they are paid by the hour, week, or month; paid a commission on sales; paid on a piecework basis; paid partially in tips; or paid a set fee for the work.

> TIP
> **Some businesses are exempt from the FLSA.** A handful of businesses are exempt from the FLSA—for example, most small farms are not covered. It's not likely your business falls within any of these exemptions, but for details of these exemptions, check with the nearest office of the U.S. Labor Department's Wage and Hour Division.

Workers Exempt From Overtime and Minimum Wage Requirements

Several categories of workers are exempt from the FLSA, even if their employer is covered by the law and even if they themselves are employees. They can work as much overtime as they want and you won't have to pay time-and-a-half. The most common exemptions are for white collar workers and outside salespeople.

White-Collar Workers

Many white-collar workers are exempt from the FLSA. The FLSA divides such workers into three categories:

- **Executives.** Employees who manage two or more employees within a business or a department, and who can hire, fire, and promote employees.
- **Administrators.** Employees who perform specialized or technical work related to management or general business operations.
- **Professionals.** Employees who perform original and creative work or work requiring advanced knowledge normally acquired through specialized study—for example, engineers and accountants.

To be exempt from the FLSA, these employees must be paid a minimum weekly salary of $455.

RESOURCE

Get free information on overtime requirements and exemptions. At the Department of Labor's website, www.dol.gov, you can find lots of fact sheets and other information on the federal overtime rules.

Outside Salespeople

An outside salesperson is exempt from FLSA coverage if he or she regularly works away from your place of business while making sales or taking orders.

Typically, an exempt salesperson will be paid primarily through commissions and will require little or no direct supervision.

Computer Specialists

Computer system analysts and programmers whose primary duty is systems analysis, systems design, or high-level programming are exempt from the FLSA if they receive a salary of at least $455 a week or, if paid by the hour, at least $27.63 an hour.

Other Workers

Several other types of workers are exempt from the overtime pay and minimum wage provisions of the FSLA. Others are exempt only from the overtime pay rules.

Workers Exempt from FSLA Rules	
Exempt from Overtime and Minimum Wage Rules	**Exempt from Overtime Rules Only**
Volunteer workers	Rail, air, and motor carrier employees
Outside salespeople	Employees who buy poultry, eggs, cream, or milk in their unprocessed state
Employees of organized camps and religious and nonprofit educational conference centers that operate fewer than seven months a year	Those who sell cars, trucks, farm implements, trailers, boats, or aircraft
Employees of certain small newspapers and newspaper deliverers	Mechanics or parts persons who service cars, trucks, or farm implements
Workers engaged in fishing operations	Announcers, news editors, and chief engineers of certain broadcasting stations
Seamen on international vessels	Local delivery drivers or drivers' helpers who are compensated on a trip rate plan
Employees who work on small farms	Agricultural workers
Certain switchboard operators	Taxi drivers
Casual domestic babysitters and persons who provide companionship to those who are unable to care for themselves (see Chapter 7)	Movie theater employees
	Domestic service workers who live in the employer's home (see Chapter 7)

Classifying Workers Under the FLSA

If a worker does not fall into any of the exempt categories discussed above, the FLSA will apply only if the worker is an employee. The Department of Labor and most courts use an economic reality test to determine the status of workers for FLSA purposes.

The economic reality test is also used to determine employee status for the Worker Adjustment and Retraining Notification Act (which requires employers to give advance notice of plant closings and mass layoffs). It is also applied frequently by courts in determining employee status in cases involving Title VII of the Civil Rights Act of 1964, the Age Discrimination in Employment Act, and the Americans with Disabilities Act.

RESOURCE

Get free information about these and other employment laws on Nolo's website at www.nolo.com. The websites of the Department of Labor (www.dol.gov) and the Equal Employment Opportunity Commission (www.eeoc.gov) are also very informative.

Under the economic reality test, workers are employees if they are economically dependent upon the businesses for which they render services. This can be a rather difficult test to apply. After all, taken to its logical extreme, all workers could be considered employees because all, to some extent, are economically dependent on the people they work for.

As a general rule, however, the economic reality test will classify as employees all workers who would be considered employees under the common law test (see Chapter 2 for an in-depth discussion of this test). It will also classify as employees workers whom government agencies and courts feel need and deserve special protections. These are primarily low-skill, low-paid workers—the type of workers labor and workers' compensation laws were originally intended to help.

This is borne out by the type of factors courts examine to gauge the degree of a worker's dependence on a hiring firm. They include:

- whether the worker has the right to control how to do the job
- whether the worker has an opportunity for profit or loss depending on the worker's skill
- whether the worker has invested in equipment or materials or hired helpers
- whether the worker's job requires special skills
- whether the work relationship is permanent, and
- whether the worker's job constitutes an essential part of the hiring firm's business.

Highly skilled, highly paid workers with substantial investments in tools and equipment are likely to be considered ICs under this test as long as they don't work full time for just one firm. In contrast, a worker who doesn't earn much, has low skills, and doesn't have to use much individual initiative to earn a living will probably be an employee.

A few court decisions help illustrate what a judge may emphasize when applying the economic reality test.

In one case, a natural gas pipeline construction company hired pipe welders and classified them as ICs. Twenty of them sued the company, claiming they were entitled to overtime pay because they were really employees. The court concluded that the workers were ICs under the economic reality test and therefore not entitled to overtime pay. The court noted that:

- The welders' jobs were highly specialized and required great skill.
- The welders moved from company to company and from job to job, usually working no more than six weeks at a time for any one company.
- The company exercised no control over how the welders did their jobs. Instead, the company's customers specified the type of welding procedures to be used and then tested the finished results.
- The welders owned all their own welding equipment and trucks, with an average cost of $15,000.
- The welders' success depended on using their initiative to find consistent work by moving from job to job.
- Although they were paid an hourly rate, the welders' opportunity for profit or loss depended mostly on their abilities to find work and minimize welding costs.

Based upon these facts, the court concluded that the welders were ICs. (*Carrell v. Sunland Const. Inc.*, 998 F.2d 330 (5th Cir. 1993).)

In another case, the Department of Labor claimed that a nightclub operator had incorrectly classified topless dancers as ICs and was liable for overtime pay and for failing to pay the minimum wage. The court agreed. Even though the dancers' compensation was derived solely from tips they received from customers, the court found they were employees under the economic reality test. The dancers were economically dependent upon the nightclub because it set their work schedules and the minimum amounts they could charge for table dances and couch dances. Moreover, the club played the major role in attracting customers through advertising, providing customers with food and beverages, and other means. The only initiative the dancers provided was deciding what to wear and how provocatively to dance. (*Reich v. Circle C. Investments, Inc.*, 998 F.2d 324 (5th Cir. 1993).)

Avoiding Problems

The easiest way to avoid problems with overtime pay requirements is to prevent workers from putting in more than 40 hours a week. If a person is clearly an employee, you can simply prohibit him or her from working overtime. But if you classify a worker as an IC, you should not directly specify how many hours he or she should work, either orally or in a written IC agreement. Doing so makes the worker look like an employee, not only for FLSA purposes but for IRS and other purposes as well. It's really none of your business how long an IC works. You can be concerned only with the results an IC achieves, not how the worker achieves them.

Avoid giving workers who are not clearly ICs more work than they can do in a 40-hour week. This may mean you have to plan ahead so you can lengthen deadlines or hire more ICs to do the needed work.

Record-Keeping Requirements

The FLSA requires you to keep records of wages and hours for employees. You do not have to keep such records for independent contractors.

Federal Labor Relations Laws

The National Labor Relations Act, or NLRA (29 U.S.C. § 151 and following), gives most employees the right to unionize. This enables them to negotiate collective employment contracts through union representatives rather than having to deal with employers individually.

The National Labor Relations Board (NLRB) administers the law and interprets its provisions. The NLRB conducts union elections and enforces the NLRA's rules of conduct, determining whether employers have engaged in unfair labor practices.

ICs Are Not Covered

The NLRA applies only to employees. ICs have the right to form a union if they wish to do so, but they are not protected by the NLRA. You can decline to use the services of ICs who form a union or simply express support for a union. You can't do this with employees who are covered by the NLRA.

Employees Exempt From the NLRA

Not all private sector employees are covered by the NLRA. Exempt employees include:
- managers and supervisors
- confidential employees—such as company accountants
- farmworkers
- members of an employer's family
- most domestic workers, and
- workers in certain industries—such as the railroad industry—that are covered by other labor laws.

Determining Worker Status

If a worker does not fall within one of the classes of NLRA-exempt employees, you need to decide whether he or she is an employee or IC for NLRA purposes. The common law right-of-control test is used for this purpose. (See Chapter 2 for a discussion of the common law test.)

Family and Medical Leave Act

The Family and Medical Leave Act (FMLA) requires employers with 50 or more employees to give eligible employees up to 12 work weeks of leave during any 12-month period for the following reasons:
- to care for a newborn or newly adopted child
- to care for a seriously ill spouse, child, or parent, or
- to recuperate from their own serious health condition.

(29 U.S.C. §§ 2601–2654.)

To qualify for leave, an employee must have worked for the employer for at least 12 months and for at least 1,250 hours during the previous 12 months. The employer is not required to pay the employee during the leave period.

ICs are not protected by the FMLA. The Department of Labor enforces the FMLA and uses the economic reality test to determine whether workers are employees or ICs. (See above for information on the economic reality test.)

Fair Credit Reporting Act

The Fair Credit Reporting Act (FCRA; 15 U.S.C. § 1681 and following) imposes a number of legal obligations on consumer reporting agencies, such as credit bureaus, and on those who use credit information and other information gathered by third parties to make important decisions. In part, it governs the extent to which employers can obtain credit reports and background information (including criminal records) from consumer reporting agencies. The FCRA applies to firms that hire ICs as well as those that hire employees.

The FCRA places the following requirements on businesses that hire employees and ICs:

- **Before obtaining a report:** Before obtaining a credit report or similar background report, the business must tell the applicant that it might use information in the report for decisions related to his or her employment (or his or her work as an IC) and get their written permission to do so.
- **Before taking adverse action:** Before taking any adverse action (such as denying, suspending, reassigning, or terminating the person) based in whole or in part on a report, the business must send a notification letter to the job applicant or current employee or IC. The notification letter must include a copy of the report and a description of the rights provided by the FCRA.
- **After taking adverse action:** Once a business decides to take an adverse action based in whole or in part on a consumer report, it

must provide the job applicant or current employee or IC notice containing:

- the consumer reporting company's contact information
- a statement that the company that supplied the report didn't make the decision to take the unfavorable action and can't give specific reasons for it, and
- notice of the right to dispute the accuracy or completeness of the report, and to get an additional free report from the company by requesting it within 60 days.

For more details, see "Using Consumer Reports: What Employers Need to Know" (available at www.business.ftc.gov/documents/bus08-using-consumer-reports-what-employers-need-know).

The FCRA is enforced by the Consumer Financial Protection Bureau (www.consumerfinance.gov). Individuals can also file lawsuits in federal court against businesses that violate its terms. The states and several other federal agencies and departments also have the power to enforce this law.

Antidiscrimination Laws

The federal government and most states have laws prohibiting discrimination in the workplace. Most of these laws apply only to employees, not ICs.

Federal Antidiscrimination Laws

The main federal law barring workplace discrimination is Title VII of the federal Civil Rights Act of 1964. Title VII applies to businesses that have 15 or more full-time or part-time employees. It outlaws discrimination in employment based on race, color, religion, gender, or national origin. Harassment in the workplace, including sexual harassment, is also prohibited as a variety of illegal discrimination.

Other federal laws barring workplace discrimination include:

- the Age Discrimination in Employment Act, which prohibits age discrimination against people who are 40 or more years old and applies to employers with 20 or more employees

- the Pregnancy Discrimination Act, which bars employers from discriminating against employees based on pregnancy, birth, or related conditions and applies to employers with 15 or more employees
- the Immigration Reform and Control Act, which makes it illegal for all employers to discriminate against people who aren't U.S. citizens but who have been legally admitted to the United States
- the Equal Pay Act, which requires all employers to provide equal pay and benefits to men and women who do the same job or jobs requiring equal skill, effort, and responsibility, and
- the Americans with Disabilities Act, which protects disabled people from employment discrimination and applies to employers with 15 or more employees.

With one very narrow exception (see "Beware of Local Twists," below), none of these antidiscrimination laws applies to ICs. An IC has no legal right to bring a lawsuit against you claiming that you have discriminated in violation of these statutes. And the federal agencies charged with enforcing these laws, such as the Equal Employment Opportunity Commission, have no power to handle claims brought by ICs.

But this does not mean that you are off the hook as far as ICs are concerned. There are other federal and state laws that prohibit various forms of discrimination, and these laws may apply to your relationship with the IC. For example, in one case, a court held that an IC can sue a hiring company for damages for discrimination under a federal law that bars racial discrimination in private contracts—42 U.S.C. § 1981. In this case, an IC named Benjamin Guiliani, a Mexican-American man, operated a company called Danco that maintained parking lots. Danco signed a contract with a Wal-Mart store agreeing to maintain the store's parking lot. After Guiliani began working at the Wal-Mart location, he experienced hostility based on his race. Among other incidents, someone painted the words "white supremacy" on the pavement near where Guiliani unloaded his equipment, and store supervisors made derogatory comments about Latinos in front of him. Guiliani complained about the treatment, and Wal-Mart canceled the contract. Guiliani sued and won a $300,000 judgment. (*Danco, Inc. v. Wal-Mart Stores, Inc.,* 178 F.3d 8 (1st Cir. 1999).) However, courts in some other parts of the country have not extended this federal statute to ICs. For example, a Colorado court held that ICs who couldn't bring discrimination claims under Title VII couldn't bring identical claims under Section 1981. (*Lufti v. Brighton Cmty. Hosp. Ass'n,* 40 p. 3d 51 (Colo. Ct. App. 2001).)

In addition, independent contractors who have incorporated their businesses cannot personally sue other companies for race discrimination in contracting. In *Domino's Pizza v. McDonald,* 546 U.S. 470 (2006), the U.S. Supreme Court rejected a black contractor's Section 1981 claim that Domino's Pizza had broken several contracts with his corporation because of his race. Although the contractor was the sole shareholder of his corporation and entered into the contracts on its behalf, the Court found that he had no right to sue because the contracts were between Domino's and his corporation, not him personally.

Workers you've classified as ICs might be able to sue you for workplace discrimination by claiming that they should have been classified as employees. If such a worker could convince a court he or she was improperly classified, the antidiscrimination laws would then apply. A court or federal agency would likely bend over backwards to find an employment relationship if you have engaged in blatant discrimination.

In addition, an IC might be able to sue you under state antidiscrimination laws, and you might be subject to administrative action by a state antidiscrimination agency. Some of these state laws may apply to ICs. (See below.)

Until recently, courts almost always used the economic reality test to determine whether a worker was an employee or IC for purposes of federal antidiscrimination statutes. However, the trend lately is to use the common law right-of-control test, as the Supreme Court did in 2003 in a case involving the Americans with Disabilities Act. (*Clackamas Gastroenterology Associates, P.C. v. Wells,* 538 U.S. 440 (2003).) (See Chapter 2 for a discussion of this test.)

> **Beware of Local Twists**
>
> Federal courts in some parts of the country—the western United States, for example—have found that Title VII can apply to an IC if discrimination against the IC results in damage to his or her job opportunities.
>
> For example, a doctor was permitted to bring a Title VII action alleging discrimination on the basis of national origin. The doctor, clearly an IC, had submitted a bid to run a hospital's emergency room. The doctor claimed that the bid was rejected because he was Hispanic and that the rejection had adversely affected his job opportunities. (*Gomez v. Alexian Bros. Hosp.*, 698 F.2d 1019 (9th Cir. 1983).)

State Antidiscrimination Laws

Virtually every state has its own antidiscrimination law. And many states and localities prohibit forms of discrimination that aren't covered by federal law—for example, discrimination based on marital status or sexual orientation.

These laws may be enforced by a special state administrative agency, the state labor department, or state attorney general. Covered workers can also bring lawsuits alleging job discrimination against employers in state court.

Most of these state laws apply only to employees, not ICs. Some state courts use the economic reality test to determine worker status under these laws, while others use the common law test. Regardless of what test is used, state agencies and courts often take a very broad view of who qualifies as an employee under these antidiscrimination laws. It's possible, therefore, that a worker might be viewed as an IC under federal antidiscrimination laws, but as an employee under a similar state law.

Beware that the civil rights laws of a few states—Louisiana, North Dakota, and Vermont, for example—might include ICs as well as employees. And if your state's constitution prohibits discrimination, that provision may protect ICs as well.

For more information, contact your state labor department. (See www. dol.gov/dol/location.htm.)

Worker Safety Laws

The federal Occupational Safety and Health Act (OSH Act) requires employers to keep their workplaces safe and free from recognized hazards that are likely to cause death or serious harm to employees. (29 U.S.C. §§ 651 to 678.) Employers must also provide safety training to employees, inform them about hazardous chemicals, notify government administrators about serious workplace accidents, and keep detailed safety records.

The OSH Act applies to businesses that affect interstate commerce. The legal definition of interstate commerce is so broad that almost all businesses are covered.

The OSH Act is enforced by the federal Occupational Safety and Health Administration, or OSHA, a unit of the Department of Labor. OSHA can impose heavy penalties for legal violations and can set additional workplace standards.

OSHA Coverage of ICs

The OSH Act applies only to employees, not to ICs. OSHA uses the economic reality test to determine whether workers are employees or ICs. (See above for an in-depth discussion of this test.) OSHA has interpreted the test broadly to bring as many people as possible within the coverage of the law, making a lot of people employees who might not be under other tests, such as applicants for employment. (29 C.F.R. § 1977.5(b).)

The training and record-keeping requirements mentioned above don't apply to ICs. In addition, ICs do not have the legal right to complain to OSHA about safety violations, nor can they refuse to work if such violations persist. However, OSHA regulations requiring employers to notify workers about hazardous chemicals appear to apply to ICs as well as to employees. (29 C.F.R. § 1910.1200(c).)

Importance of Maintaining a Safe Workplace

Even though OSHA cannot impose penalties against you if you have no employees, it's important for you to maintain a safe workplace. ICs who perform services at your workplace may be able to sue you for negligence and obtain monetary damages if they are injured because of hazardous or unsafe conditions.

Immigration Laws

Some workers in the United States are immigrants. And some of these immigrants work illegally—that is, they are not U.S. citizens and don't have a green card or other documentation of their legal status.

All employers must verify that their employees are either U.S. citizens or nationals, or legal aliens authorized to work in the United States. For details, refer to the United States Citizenship and Immigration Services website at www.uscis.gov.

You are not required to verify citizenship when you hire an IC. The government uses the common law right-of-control test to determine whether a worker is an employee or IC for immigration purposes. (See Chapter 2 for an in-depth discussion of the common law test.)

Although you are not required to verify the immigration status of ICs or others coming within these exceptions, it is still illegal for you to hire any worker whom you know to be an illegal alien. The federal government can impose a fine of up to $2,000 for the first offense.

Intellectual Property Ownership

What Is Intellectual Property? 232

Laws Protecting Intellectual Property 232

    Copyright Law 232

    Patent Law 233

    Trade Secret Law 234

Copyright Ownership 234

    Works Made for Hire 235

    IC Works That Are Not Made for Hire 238

    Determining Whether Workers Are Employees or ICs 240

Trade Secret and Patent Ownership 241

    Inventions and Trade Secrets Created by Workers 241

    Revealing Trade Secrets to Third Parties 242

This chapter explains the rights and responsibilities of those who hire ICs to help create intellectual property. This includes not only high-technology companies and publishers, but any company that has information it wants to keep from its competitors.

What Is Intellectual Property?

Intellectual property is a generic term describing products of the human intellect that have economic value. It includes works of authorship (such as writings, films, and music), inventions, and information or know-how not generally known.

Intellectual property is considered property because the law gives the owners of such works legal rights similar to the rights of owners of real estate or tangible personal property such as automobiles. Intellectual property may be owned, bought, and sold the same as other personal property.

Despite these similarities, there are significant ways in which owning intellectual property is quite different from owning a house or car. For example, if you pay an IC to build a house, you own the house. But you can pay an IC to create intellectual property and yet not own the finished product.

Laws Protecting Intellectual Property

There are three separate bodies of law that protect most types of intellectual property: copyright, patent, and trade secret law.

Copyright Law

The federal copyright law (17 U.S.C. § 101 and following) protects all original works of authorship. A work of authorship is any work created by a human being that other humans can understand or perceive, either by themselves or with the help of a machine such as a film projector or television. This includes, but is not limited to, all kinds of written works, plays, music, artwork, graphics, photos, films and videos, computer software, architectural blueprints and designs, choreography, and pantomimes.

Copyright law gives the owner of a copyright a bundle of exclusive rights over how the work may be used. These include the exclusive right to copy and distribute the protected work, to create derivative works based upon it—updated editions of a book, for example—and to display and perform it. Copyright owners typically profit from their works by selling or licensing all or some of these rights to others—publishers, for example.

 **RESOURCE**
For a detailed discussion of copyright, see *The Copyright Handbook: What Every Writer Needs to Know* and *The Public Domain*, both by Stephen Fishman (Nolo).

Patent Law

The federal patent law (35 U.S.C. § 100 and following) protects inventions. To obtain a patent, an inventor must file an application with the U.S. Patent and Trademark Office in Washington, DC. If the Patent Office determines that the invention meets the legal requirements, it will issue a patent to the inventor. A patent gives an inventor a monopoly to use and commercially profit from the invention for 20 years. Anyone who wants to use or sell the invention during that time must obtain the patent owner's permission. A patent may protect the functional features of a machine, process, manufactured item, or composition of matter, or the ornamental design of a nonfunctional feature. A patent also protects improvements of any such items.

 RESOURCE
Need more information on patents? Nolo has lots of great resources on patents, including:
- *Patent it Yourself,* by David Pressman, updated by Thomas J. Tuytschaevers, Esq. and
- *Patent Pending in 24 Hours,* by Richard Stim and David Pressman.

Trade Secret Law

A trade secret is information or know-how that is not generally known by others and that provides its owner with a competitive advantage in the marketplace. The information can be an idea, written words, a formula, process or procedure, a technical design, a customer list, a marketing plan, or any other secret that gives the owner an economic advantage.

If a trade secret owner takes reasonable steps to keep the confidential information or know-how secret—for example, does not publish it or otherwise make it freely available to the public—the laws of most states will protect the owner from disclosures of the secret by:

- the owner's employees
- people who agree not to disclose it
- industrial spies, and
- competitors who wrongfully acquire the information.

For detailed information, see *Patent, Copyright & Trademark: An Intellectual Property Desk Reference*, by Richard Stim (Nolo).

Copyright Ownership

A work of authorship is automatically protected by copyright from the moment it is created. At that same moment, someone becomes the owner of the copyright. If you pay an IC to create a copyrightable work on your behalf, you probably want to be the copyright owner, so you have the exclusive right to copy, distribute, and otherwise economically exploit the work. Without these rights, your ability to use the work will be very limited, even though you paid for it.

There are two ownership possibilities. Either:

- the work will be made for hire, in which case you will automatically be the copyright owner, or
- the work will not be made for hire, in which case the IC will own the copyright unless he or she signs an agreement transferring those rights to you.

Failing to Obtain Copyright Transfers From ICs

If you fail to obtain a copyright transfer from an IC, the best thing that can happen is that you will be considered a coauthor of the work the IC helps create. For this to occur, you or somebody who works for you must actually help the IC create the work. Giving suggestions or supervision is not enough to be a coauthor.

If you qualify as a coauthor, you and the IC will jointly share copyright ownership in the work. As a coauthor, you're entitled to use or let other people use the work without obtaining approval of the other coauthor. But any profits you make must be shared with the other coauthor or coauthors.

If you don't qualify as a coauthor, at most you will have a nonexclusive right to use the work. For example, if you hired an IC to create a computer program, you will be able to use the program without asking the IC for permission. But you won't be allowed to sell or license any copyright rights in the work because you won't own any. The IC will own all the rights and will be able to sell or license them without your permission and without sharing the profits with you.

Works Made for Hire

When you pay someone to create a work made for hire, you automatically own all the copyright rights in the work. Indeed, you are considered the work's author for copyright purposes, even though you didn't create it. The actual creator of a work made for hire has no copyright rights at all. All the creator receives is whatever compensation you give him or her.

As the "author," you own all the exclusive rights that make up a copyright, such as the right to copy and distribute the work. You can exercise these rights yourself, sell or license them to others, or do whatever else you want with them. The person or people you paid to create the work have no say over what you do with your copyright rights in the work.

There are two types of works made for hire. They include:

- works created by employees within the scope of their employment, and
- certain types of specially commissioned works created by ICs.

Works by Employees

All works of authorship your employees create within the scope of their employment are works made for hire. This means you automatically own all the copyright rights in such works. You aren't legally required to have your employees sign agreements relinquishing their copyright rights in works made for hire. However, costly disputes can develop concerning whether a work is created within the scope of employment. For example, if an employee creates a work partly at home outside working hours, he or she might claim it is not a work for hire because the work was done outside the scope of employment.

For this reason, it is a very good idea to have a written agreement describing the employee's job duties so it will be clear whether a work is created within the scope of employment. It's also wise to include a provision assigning or transferring to you the copyright rights in any job-related work that does not qualify as a work made for hire.

Specially Commissioned Works by ICs

Certain types of specially commissioned or ordered works created by ICs are also considered to be works made for hire in which the hiring firm automatically owns all copyright rights. However, you and the IC must both sign an agreement stating that the work is made for hire. (See Chapter 12 for more information about agreements with ICs.)

Nine categories of works can be IC-created works made for hire. They are:

- a contribution to a collective work—a work created by more than one author, such as a newspaper or magazine or an anthology or encyclopedia
- a part of an audiovisual work—for example, a motion picture screenplay
- a translation
- supplementary works—for example, forewords, afterwords, supplemental pictorial illustrations, maps, charts, editorial notes, bibliographies, appendixes, and indexes
- a compilation—for example, an electronic database
- an instructional text
- a test
- answer material for a test, and
- an atlas.

Special Rules for California

Under California law, a person who commissions a work made for hire is considered to be the employer of the creator of the work for purposes of workers' compensation, unemployment insurance, and unemployment disability insurance laws. (Cal. Lab. Code § 3351.5(c); Cal. Unemp. Ins. Code §§ 621 and 686.)

No one is sure what impact this has on those who commission works made for hire in California. Neither the California courts nor state agencies have addressed the question. However, it may mean that the hiring firm has to obtain workers' compensation and unemployment insurance coverage for the person who created the work. It might also mean that special penalties could be assessed against a hiring firm that does not pay the creator money due after he or she is discharged or resigns.

In addition, the IRS could use these California laws as an excuse to classify creators as employees for federal tax purposes.

These potential requirements and liabilities are good reasons why it might be desirable for those commissioning work in California not to enter into work-made-for-hire agreements, but instead have the creator assign the desired copyright rights to the hiring firm in advance. (See below.)

However, there is one legal drawback to using assignments of copyright instead of work-made-for-hire agreements with ICs: Under the federal copyright law, the IC-author of a work that is assigned has the option of terminating the assignment starting 35 years after the date of the assignment, and continuing for a period of five years. The author can thereby get all the copyright rights back without paying anything to the hiring firm. Few works have a useful economic life of more than 35 years, so this termination is usually not a problem. But, it could be a problem for works whose exploitation is expected to be long-lived. In this event, it may be wiser to use a work-made-for-hire agreement and classify the author as an employee.

These California laws apply only to contracts between hiring firms and individual ICs—they don't apply when firms hire corporations, LLCs, or other business entities. So, another way to avoid the problem is to deal with an IC who operates through a business entity instead of as an individual. Make sure the work-made-for-hire agreement is entered into with, and signed by, the entity, not the individual IC. (See Chapter 12.)

EXAMPLE: The editor of *The Egoist Magazine* asks Gloria, a freelance writer, if she would be interested in writing an article for the magazine on night life in Palm Beach. Gloria agrees and the editor sends her an agreement to sign setting forth such terms as compensation, the deadline for the article, and the article's length. The agreement also states that the article "shall be a work made for hire." Gloria signs the agreement, writes the article, and is paid by the magazine. Because the article qualifies as a work made for hire, the magazine is the initial owner of all the copyright rights in the article.

Gloria owns no copyright rights in the article. As the copyright owner, the magazine is free to sell reprint rights in the article and to sell film and television rights, translation rights, and any other rights anyone wants to buy. Gloria is not entitled to license or sell any rights in the article because she doesn't own any; she gave up all her copyright rights by signing the work-made-for-hire agreement.

IC Works That Are Not Made for Hire

Works of authorship created by ICs that do not fall within the list of nine specially commissioned works discussed above can never be works made for hire. This means that the IC, not the hiring firm, initially owns the copyright in such a work. As the copyright owner, the IC has the exclusive right to copy, distribute, and create new works based on the work. Even though you paid the IC to create the work, you won't own any of these exclusive rights. You may end up with only a limited right to use the work.

EXAMPLE: Tom hires Jane, a freelance programmer, to create a computer program. Tom and Jane have an oral work agreement and Jane qualifies as an IC. She works at home under her own direction, sets her own hours, and uses her own computer. Jane completes her work and delivers her code, and Tom pays her.

The program is not a work made for hire because Jane is an IC, not Tom's employee, and a computer program does not fall within one of the categories of works created by ICs that can be works made for hire. And, in any event, Jane never signed a work-made-for-hire agreement. This means that Jane owns all the copyright rights in the program. As the copyright owner, she has the exclusive right to sell the program to others or permit them to use it. Even though Tom paid Jane to create the program, he doesn't own it and can't sell or license it to others.

Fortunately, it's easy to avoid this unhappy result. Simply require all ICs who create copyrightable works for you to sign written agreements assigning you the copyright rights you need before they begin work on a project.

An assignment is simply a transfer of copyright ownership. You can obtain all the copyright rights in the work, or part of them. It's up to the IC and you to decide which rights to transfer. As discussed above, a copyright is really a number of rights including the exclusive rights to copy, distribute, perform, display, and create derivative works from a work. Each of these rights can be sold or licensed together or separately. They can also be divided and subdivided by geography, time, market segment, or in any other way you can think up. For example, you could obtain the right to copy and distribute a work in North America for ten years.

An assignment can be made either before or after a work is created, but must be in writing to be valid. (See Chapter 12 for information on creating an assignment.)

> EXAMPLE: Tom hires Jane, a freelance programmer, to create a computer program. Before Jane starts work, Tom has her sign an independent contractor agreement providing in part that she transfers all her copyright rights in the program to Tom. Jane completes her work and delivers the program, and Tom pays her. Tom owns all the copyright rights in the program.

Obtaining copyright ownership through an assignment is not legally the same as owning a work made for hire. When you own a work made for hire you are considered the work's author, even though you didn't create it. You automatically own all the copyright rights in the work. You are not considered the author when you obtain a copyright through an assignment, and you acquire only those rights covered by the assignment.

You'll usually want the assignment to transfer all of the IC's copyright rights. When you do this, the only practical difference between an assignment and a work made for hire is that the IC or his or her heirs can terminate the assignment 35 to 40 years after it was made. However, because very few works have a useful economic life of more than 35 years, this right doesn't add up to much.

Determining Whether Workers Are Employees or ICs

It should be clear by now that it is very important to know whether any person you hire to create a work of authorship qualifies as an employee or IC for copyright ownership purposes. You automatically own the copyright in works created by employees within the scope of employment, but this is emphatically not the case with works created by ICs.

The common law right-of-control test is used to determine whether a worker is an IC or employee for copyright purposes. (See Chapter 2 for an in-depth discussion of the common law test.)

> **EXAMPLE:** Marco, a professional photographer, took photographs for several issues of *Accent Magazine*, a trade journal for the jewelry industry, over a six-month period. Marco had an oral agreement with the magazine and was paid a fee of about $150 per photograph. Marco made no agreement with the magazine concerning copyright ownership of the photos. Marco, who had not signed a work-made-for-hire agreement, claimed that he owned all the copyright rights in the photos.
>
> The court concluded that Marco was an IC. Marco was an experienced and skilled photographer. He used his own equipment and worked at his own studio, on days and times of his choosing, without photography assistants hired by the magazine. No income tax was withheld from his payments and he received no employee benefits. He performed discrete assignments for the magazine, rather than hourly or periodic work. Because Marco owned the copyright in the photos, the court held that the magazine had to pay him a licensing fee when it reused them. (*Marco v. Accent Publishing Co., Inc.*, 969 F.2d 1547 (3d Cir. 1992).)

You Can't Have It Both Ways

Courts don't look favorably on businesses that don't treat workers even-handedly for copyright ownership purposes. In one case, for example, a federal court held that a part-time programmer employed by a swimming pool retailer was not the company's employee for copyright purposes; the programmer was therefore entitled to ownership of a program he wrote for the company. The court stated that the company's failure to provide the programmer with health, unemployment, or life insurance benefits, or to withhold Social Security, federal, or state taxes from his pay was a virtual admission that he was an independent contractor. The court stressed that the company could not treat the programmer as an independent contractor for tax purposes and then turn around and claim he was an employee for copyright ownership purposes. (*Aymes v. Bonelli*, 980 F.2d 857 (2d Cir. 1992).)

The moral is that if you treat a worker as an IC for IRS purposes, you had better assume he or she is an IC for copyright ownership purposes as well.

Trade Secret and Patent Ownership

The rules for determining ownership of trade secrets and patentable inventions by ICs are essentially the same.

Inventions and Trade Secrets Created by Workers

Whenever you hire any worker to create or contribute to the creation of a patentable invention or information you wish to maintain as a trade secret, you must have the worker sign an agreement transferring his or her ownership rights to your company. This is so whether the worker is an employee or IC.

Such an intellectual property ownership transfer is called an assignment. Assignments are common practice among high-technology firms and other businesses that create patentable inventions or valuable trade secrets. (See Chapter 12 for information about creating an assignment.)

If you don't have a signed assignment before work begins, you can still obtain ownership of any inventions or trade secrets an IC creates on your behalf, but you may be in for a costly legal dispute. You'll have to prove that the worker was hired to develop a specific product or to help create inventions for you.

Revealing Trade Secrets to Third Parties

When you hire ICs to perform services for you, it is sometimes necessary for you to give them access to sensitive business information that you don't want your competitors to know. For example, it may be necessary to reveal highly valuable customer lists to an IC salesperson.

Even in the absence of a written agreement saying so, ICs probably have a duty to keep such information confidential. But just to make sure, it's wise to include a confidentiality clause in an IC agreement, requiring the IC to keep your proprietary information confidential. (See Chapter 12 for information about confidentiality clauses.)

Strategies for Avoiding Trouble

Hiring Incorporated Independent Contractors...244

    Sole Proprietorships...244

    Partnerships...246

    Limited Liability Companies...247

    Corporations..248

Employee Leasing...252

    Leased Workers' Employment Status...253

    Dealing With Established Leasing Companies...259

    Using Written Agreements..259

This chapter explains a few strategies for avoiding trouble when hiring ICs. If you follow these rules, you'll reduce your chances of getting audited—and improve your odds of winning any audit you do have to face. These methods have been known to legal and employment professionals for years. But you don't have to hire a highly paid expert to use them.

Hiring Incorporated Independent Contractors

The single most effective thing you can do to avoid IRS and other government audits is to hire ICs who have incorporated their own businesses, rather than those who operate as sole proprietors or partnerships.

To understand why this works, you need to know a little about the various legal forms a business can take. ICs can legally operate their businesses as:

- sole proprietorships
- partnerships
- limited liability companies, or
- corporations.

> CAUTION
>
> **There is nothing in a name.** ICs may call themselves by a variety of names: consultants, independent businesspeople, freelancers, self-employed workers, entrepreneurs, and the like. None of these names has any legal significance, however. ICs can use any of them no matter how their businesses are organized. What's important is whether their businesses are sole proprietorships, partnerships, limited liability companies, or corporations.

Sole Proprietorships

A sole proprietorship is simply a one-owner business. Any person who starts a business and does not incorporate or have a partner is automatically a sole proprietor.

A sole proprietor is neither an employee nor an IC of the proprietorship. The owner and the sole proprietorship are treated as a single entity for tax purposes. The business does not pay FICA or FUTA taxes on the owner's income or withhold income tax. Instead, the sole proprietor reports business income and losses on his or her own individual federal tax return, Form 1040, Schedule C. Sole proprietors must pay all their FICA taxes themselves in the form of self-employment taxes. These taxes are reported on Schedule SE.

> **EXAMPLE:** Imelda operates a computer consulting business as a sole proprietorship. She is the sole owner of the business. For tax purposes, Imelda and her proprietorship are one and the same. She must report all the income she receives from her clients on her individual Form 1040, Schedule C. She does not file a separate tax return for her business.

A sole proprietorship is by far the simplest and easiest way to legally operate a business, and it costs virtually nothing to start. For this reason, many ICs are sole proprietors.

Many sole proprietors qualify as ICs. You likely won't have problems proving IC status if a sole proprietor is clearly running an independent business—for example, offers services to the public, has multiple clients, incurs substantial ongoing business expenses, such as workplace rental and insurance, is paid by the project, and hires and pays assistants.

But in borderline cases, the fact that a worker is a sole proprietor won't help you in an audit. Sole proprietors who don't hire assistants can look a lot like employees. They're working on their own, just like employees. They're selling their personal services to you, just like employees do. You pay them directly, just as you do employees. They may deposit the money in a personal account, just like employees do. And like employees, they don't have corporate meetings, partnership agreements, or other formalities proving that they are running a separate business.

The bottom line is that an IRS or other government auditor is more likely to question the status of a sole proprietor you've hired than a corporation or partnership. Unfortunately, you may not be able to avoid hiring a sole proprietor, because this is how most ICs do business.

Partnerships

A partnership is formed automatically whenever two or more people go into business together and do not form a corporation or a limited liability company. This form of business is similar to a sole proprietorship, except there are two or more owners. Like a sole proprietorship, a partnership is legally inseparable from the owners—the partners.

Partners share in profits or losses in the manner in which they've agreed. The partnership itself does not pay taxes, although it must file an annual tax form. Instead, partnership income and losses are passed through the partnership directly to the partners, who report them on their individual federal tax returns, Form 1040, Schedule E.

Partners are neither employees nor ICs of their partnership; they are self-employed business owners. A partnership does not pay FICA and FUTA taxes on the partners' income, nor does it withhold income tax. Like sole proprietors, partners pay their own taxes.

> **EXAMPLE:** Brenda, Dave, and Mike start their own computer consulting business. They form a partnership in which all three are partners. For tax purposes, their lives are pretty much the same as if they were sole proprietors. Each partner must pay his or her own FICA and FUTA taxes, and each must report his or her share of partnership income and losses on an individual tax return. Brenda, Dave, and Mike are not employees of their partnership; each is a business owner.

Relatively few ICs do business as partnerships. But if you have the choice between hiring a sole proprietor and a partnership, you're usually better off hiring a partnership. Partnerships simply look more like independent businesses than most sole proprietorships. Partnerships involve two or more people in business together, not a single person. Partners' relationships with each other are governed by state partnership laws and partnership agreements, which can be complex. Also, partnerships can have their own bank accounts and own property. This means you can pay the partnership for the work, rather than paying the partners directly.

However, the IRS can still decide that partners are really employees, even if you pay the partnership itself rather than individual partners.

Limited Liability Companies

The limited liability company (LLC) is one of the most popular forms of business in the United States. An LLC is essentially a combination of a partnership/sole proprietorship and a corporation. Although the LLC provides the limited liability of a corporation, it is taxed as either a partnership or a sole proprietorship. An LLC is easy to form and to run. For example, unlike a corporation, LLCs do not have to hold regular ownership and management meetings.

Ordinarily, the LLC itself does not pay taxes. Rather, all profits and losses pass through the LLC and are reported on the owners' (or members') individual tax returns. If the LLC has only one member, the IRS treats it as a sole proprietorship for tax purposes. If the LLC has two or more members, the IRS treats it as a partnership for tax purposes. Like a partnership or corporation, an LLC can have its own bank accounts and own title to property.

RESOURCE

For a detailed discussion of LLCs, see *Form Your Own Limited Liability Company,* by Anthony Mancuso (Nolo).

A business seeking to avoid worker classification audits will probably be better off hiring an IC who has formed an LLC than one who works as a sole proprietor. Forming an LLC helps show the IC is running an independent business.

However, it isn't a bulletproof strategy. Payments to LLCs must be reported to the IRS on Form 1099-MISC, unless the LLC elects to be treated as a corporation for tax reporting purposes (few LLCs do this). Form 1099 is an important audit lead for the IRS—a lead the agency does not have when you hire a corporation, because you do not have to file a Form 1099 when you pay a corporation. (See Chapter 11 for more about paying and filing IRS forms for ICs.)

Another drawback to hiring LLCs rather than corporations is that LLC owners ordinarily are not employees of the LLC for tax purposes. They are business owners. The LLC is not required to withhold and pay employment taxes for them. In a small corporation, the owners are usually corporate employees.

Moreover, the informality of an LLC won't help you when dealing with the IRS or other government agencies. A corporation that keeps minutes, holds required meetings, and otherwise observes business formalities looks a lot more like an independent company than an LLC does.

Corporations

Business owners create a corporation by filing articles of incorporation with the appropriate state agency—usually the secretary of state or corporations commissioner. Once this is done and the appropriate fees are paid, the corporation becomes a separate legal entity—distinct from its owners, the shareholders. It can hold title to property, sue and be sued, have bank accounts, borrow money, and hire employees and ICs.

Corporate Officers, Directors, and Shareholders

In theory, a corporation is made up of three groups:
- those who direct the business—called directors
- those who run the business—called officers, and
- those who invest in it—called shareholders.

In the case of small business corporations, however, all of these roles are often filled by one person.

All corporate officers—the president, vice president, treasurer, and any others—are automatically considered corporate employees. Corporate shareholders who are not officers but perform full- or part-time services for the corporation are also employees of the corporation. The corporation must pay employment taxes for its employees.

Usually, the owners of an incorporated small business are employees of the corporation, because they serve as officers or perform services for the corporation. The corporation must deduct federal income tax withholding from the owners' wages and pay employment taxes and state payroll taxes as well.

EXAMPLE: Suzy has been operating a one-person sales and marketing business as a sole proprietor. When she incorporates the business, Suzy is the sole shareholder, director, and president of Suzy's Sales Services, Inc. She is no longer self-employed in the eyes of the tax law. Instead, she is a full-time employee of her corporation. As an employee, she earns wages from her corporation, just as if she didn't own the business. The corporation must withhold Suzy's federal income taxes, pay her FUTA and half her FICA taxes, and also pay state payroll taxes.

Benefits of Hiring Incorporated Workers

When you hire incorporated outside workers, you enter into a three-tiered relationship. You pay the worker's corporation, which pays the worker, who is an employee of the corporation. Legally, you have no direct relationship with the worker at all—only with the worker's corporation, which cannot be classified as your employee.

EXAMPLE: Acme Widget Company hires Sam's Sales Services, Inc., to sell widgets. Acme pays no money to Sam directly, even though he is the sole shareholder and president of Sam's Sales Services, Inc., and is the person doing all the work. Instead, Acme pays Sam's corporation, which is Sam's employer. It's the corporation's responsibility to pay state and federal payroll taxes for Sam.

When you hire an incorporated worker, the corporation stands between you and the worker. Legally, the corporation is the worker's employer, not you. It is supposed to pay state and federal payroll taxes and provide workers' compensation insurance, not you. If the corporation fails to pay these taxes, IRS and state auditors will go to the corporation and its owners, not you, unless the corporation is a sham. (See below.)

Even if you're audited, you'll have an easier time proving that an incorporated worker is an IC. Forming a corporation is expensive and time-consuming, and operating one can be burdensome as well. Auditors are usually greatly impressed by the fact that a worker has gone to the time and trouble to form a corporation. This is something that only people who are running their own businesses do. And people who are running their own businesses can't be your employees.

Indeed, the IRS audit manual on worker classification provides that an incorporated worker will usually not be treated as an employee of the hiring firm, but as an employee of the worker's corporation.

> EXAMPLE: An outpatient surgery center hired two doctors to work as administrators. Both performed the same services. However, one of the doctors had formed a medical corporation of which he was an employee. The surgery center signed a written contract with the corporation, not the doctor personally; it also paid the doctor's corporation for the doctor's services.
>
> The other doctor was a sole proprietor and had no written contract with the center. The court concluded that the incorporated doctor was not an employee of the surgery center, but the unincorporated doctor was. As a result, the center had to pay substantial back taxes and penalties for the unincorporated doctor, but not for the doctor who was incorporated. (*Idaho Ambucare Center v. U.S.*, 57 F.3d 752 (9th Cir. 1995).)

Another benefit of hiring incorporated outside workers is that you don't have to file a Form 1099-MISC when you pay a corporation. This eliminates a very important IRS audit lead. (See Chapter 11 for more about filing IRS forms for ICs.)

Problems With Corporations

Hiring an incorporated worker is a great strategy for avoiding audits, but it isn't foolproof. Often, independent contractors will form corporations (usually at a client's request) and then forget about them. This isn't good enough. The IC must file yearly corporate tax returns and keep all required state filings up to date. The IRS and other agencies will look at the following factors, among others, when determining whether a corporation is bona fide:

- Does the corporation have sufficient capitalization—that is, does it have sufficient assets to meet its obligations?

- Has the corporation issued corporate stock?
- Does the corporation maintain a corporate bank account?
- Have the owners intermingled corporate and personal accounts or funds?
- Does the corporation hold itself out to the public as a corporation?
- Does the corporation maintain corporate books and records, including corporate meeting minutes and board-of-director resolutions?
- Has the corporation filed articles of incorporation with the secretary of state or other appropriate state agency?

You might still run into trouble if the outside worker's corporation fails to pay federal or state payroll taxes. If there is evidence that the corporation is not being operated as an independent business, it's quite possible that the IRS will disregard the corporation as a sham and find that the worker is really your employee.

> **EXAMPLE:** Robert Allen Smith, a Hawaiian lawyer, restructured his business in an attempt to avoid paying employment taxes for his employees. Smith incorporated his law practice and required each of his employees to form their own corporations. Smith's corporation then contracted with the workers' corporations for their services—the exact same services they performed when they were Smith's employees. Smith treated the workers as ICs and stopped paying federal and state payroll taxes for them. The IRS audited Smith's corporation and concluded that the incorporated workers were really his employees. The IRS and courts found Smith personally liable for more than $113,000 in back taxes and penalties. (*In re Smith*, 243 B.R. 89 (D. Haw. 1999) aff'd, 246 F.3d 676 (9th Cir. 2000).)

Finding Incorporated ICs

Skilled workers, such as lawyers, doctors, and accountants, often form their own corporations. You probably won't have much trouble hiring incorporated workers in these fields.

However, it's unusual for lower skilled and lower paid workers to be incorporated. For example, it's rare for a trucker or delivery person to be incorporated. So, you simply may not be able to find an incorporated worker to perform the services you need.

In addition, you may have to pay more to hire an incorporated worker because operating a corporation is more expensive and burdensome than being a sole proprietor or partner. In many states, a corporation must pay a minimum annual tax even if it didn't earn any money for the year. Corporations also have to hold meetings, keep records, and follow other formalities, all of which can take time and money. And the corporation may seek to pass some of the business costs on to you by charging more for its services.

> ⃠ CAUTION
>
> **Don't help workers incorporate.** You should not help a worker form a corporation or pay him or her to do so, because this makes the corporation look like a sham. True ICs who are in business for themselves form their corporations on their own initiative with their own money.

Employee Leasing

Instead of hiring workers directly, many companies lease or rent them from outside leasing companies. Such workers may be referred to as temporary employees, temps, contract employees, or contingent or casual workers. This chapter refers to them as leased employees.

Using leased employees is sometimes referred to as outsourcing or outside staffing. Whatever the practice is called, leasing employees has become an increasingly popular method for businesses seeking the services of outside workers.

Worker leasing arrangements take a variety of forms. For example, you may lease workers from an employment agency that locates the workers for you or already has them on staff. This is what temporary agencies do. In other cases, the leasing company may hire your employees and lease them back to you for a fee.

Employee leasing can give you many of the benefits of hiring ICs directly. You use them only when needed and then dispense with their services without going through the trauma and expense of laying off your own employees. You do not have to pay and withhold federal and state payroll taxes for leased workers or provide them with workers' compensation or employee benefits.

It can cost more to lease workers than to hire them directly because leasing companies have to pay the leased employees salaries and earn a profit. Many companies are willing to pay a higher price to obtain the services of highly trained and experienced workers who have been screened and selected by the leasing company. Also, you have reduced exposure to government audits.

Although employee leasing arrangements can work well, there are some potential pitfalls.

Leased Workers' Employment Status

The idea behind worker leasing is that the leased workers are supposed to be the leasing firm's employees, not yours. The leasing firm is responsible for supervising and controlling the worker's job performance, paying the leased workers' salaries, paying and withholding federal and state payroll taxes, paying for unemployment compensation, and providing workers' compensation coverage and any employee benefits. Ideally, all you do is pay the leasing firm a fee.

When you pay a leasing firm to provide you with workers, you have much less chance of being audited. As long as the leasing firm pays all applicable taxes, there is little likelihood that you'll have any problems with the IRS.

Alternative Work Arrangements

Besides leasing employees, there are many other work arrangements. They include:

- hiring part-time workers
- hiring short-term workers
- having workers work at home and communicate with the office via phone and computer—also known as telecommuting, and
- using seasonal workers.

The usual rules for determining employment status apply to these workers as well. The fact that workers work part time, short term, at home, or seasonally has relatively little impact on their status. If you have the right to control such workers on the job, they will be your employees.

The Problem of Joint Employment

Unfortunately, things don't always work out as planned when you use a leasing company. If you control a leased worker's performance on the job, you and the leasing company might both be considered the worker's employer. This is called joint employment. If you're a joint employer of a leased employee, you lose all the benefits of employee leasing. You have the same duties and liabilities as if you were the worker's sole employer.

EXAMPLE: The Merrill Lynch securities firm leased the services of Amarnare through an employment agency. Amarnare's pay and benefits were paid by the agency, not Merrill Lynch. Amarnare was fired after two weeks on the job and then sued Merrill Lynch, but not the employment agency, for unlawful discrimination, claiming that she was fired because of her sex and race. Merrill Lynch claimed it was not liable because it was not Amarnare's employer.

The court disagreed. It held that Merrill Lynch was Amarnare's joint employer, along with the employment agency, because it completely controlled her on the job. Merrill Lynch controlled

Amarnare's work assignments, working hours, and manner of performance; directly supervised her; and had the right to discharge her and request a replacement if it found her work unsatisfactory. (*Amarnare v. Merrill Lynch, Pierce, Fenner & Smith, Inc.*, 611 F.Supp. 344 (S.D. N.Y. 1984).)

If you're found to be a joint employer of a leased worker, you'll be liable not only for labor and antidiscrimination law violations, but also for providing the worker with unemployment insurance and workers' compensation, if the leasing company fails to provide them.

Avoiding Joint Employer Status

To avoid being a joint employer of a leased worker, you must give up all control over the worker. The leasing firm, not you, must control the leased worker's performance on the job.

To stay out of joint employer trouble, follow these guidelines:

- Don't ever deal or negotiate with a leased worker about when and where to work, working conditions, or the quality and price of the services to be provided by the worker. The leasing company should handle all these negotiations for you—that is, you tell the leasing company what you want and it tells the worker.
- The leasing company should have the sole right to determine whether to assign or reassign workers to perform needed tasks.
- The leasing company should set the rate of pay for the leased workers.
- The leasing company should pay the leased workers from its own account.
- The leasing company, not you, should have the right to hire or fire the leased workers. If the company fails to provide you with high-quality workers, don't fire the workers. Instead, hire a new leasing company.
- The leasing company should have the authority to assign or reassign a worker to other clients or customers if you feel the worker is not acceptable.

It's also very helpful if the leasing company provides its own supervisor or on-site administrator to manage and supervise the leased workers. This will significantly reduce your control over the workers and reduce the chances that you'll be tagged as a joint employer.

If you're unable or unwilling to relinquish all control over leased workers, you can still go ahead with a leasing arrangement. But be aware that you may be considered the leased workers' joint employer. You'll need to make certain that the leasing firm is paying all required payroll taxes, providing workers' compensation insurance, and not engaging in behavior that could get you sued, such as discriminating against workers on the basis of race or age.

Problems With Workers' Compensation Insurance

Another problem area when you lease workers is workers' compensation insurance. The employee leasing company, not you, is supposed to provide the leased workers with workers' compensation coverage.

In the past, some shady businesses saved money on workers' compensation insurance premiums by getting equally shady leasing firms to hire their employees, then lease those employees back to the business. The leasing firm would purchase workers' compensation insurance for the employees at a lower rate than the hiring firm because it was newly in business and few or no workers' compensation claims had been filed against it. (Insurance companies consider how many claims have been filed against a company—called the "experience modifier"—in setting workers' compensation premiums.) These leasing companies continually changed their names and formed new business entities. Each new entity would have a clean workers' compensation record and so would pay a lower workers' compensation premium.

This scam is no longer possible in most states. Most now require employee-leasing companies to use the same experience modifier as their client firms use for similar workers. To make sure that a leasing company is paying the proper workers' compensation premium, ask to see a copy of its workers' compensation policy showing the classifications and experience modifier used. If these are different from those in your own policy, question the leasing company closely.

Also, make sure your agreement with the leasing company requires it to notify you in writing if its workers' compensation insurance is canceled for any reason. You may have to pay premiums for leased employees if the leasing firm's insurance is canceled.

Problems With Retirement Plans

Using leased workers can also cause problems if you have a company retirement or profit-sharing plan for your employees. Such plans are usually "tax qualified"—that is, if the plan meets certain legal requirements, employer contributions to the plan are deductible business expenses for the employer and nontaxable benefits for the recipients until they retire and start taking money out. However, hiring leased workers can cause a plan to lose its tax-qualified status. This would be a tax disaster for everyone—the employer would no longer be able to deduct contributions and employees would have to pay tax on their benefits right away.

Tax qualification rules are pretty complicated. They are designed, in part, to prevent employers from setting up plans that benefit only the company bigwigs, while leaving most of their employees out in the cold. Under these rules, a plan may not provide greater benefits to highly compensated employees than to employees who are not highly compensated; a highly compensated employee is one who owns at least 5% of the business or earns at least $115,000 a year and ranks in the top 20% of employees by pay.

A plan must also cover a minimum number of employees. Generally, a plan must cover at least 70% of employees who are not highly compensated, or a specified percentage of such employees based on how many highly compensated employees benefit under the plan. (IRC § 410(b)(1).) In addition, each plan must benefit at least 40% of the company's employees, or 50 employees total, whichever is less. (IRC § 401(a)(26).)

You must treat a leased employee as your own employee for purposes of running these numbers if the employee:

- provides services under an agreement between you and a leasing organization
- has performed services for you full time for at least one year, and
- performs services under your primary direction or control.

If leased employees qualify as your employees under this test, your plan may no longer be tax qualified.

EXAMPLE: Acme, Inc., has 20 employees; 15 are covered under Acme's retirement plan. Acme contracts with Lease Co. to lease 20 workers to perform services for Acme full time. If the 20 leased employees work for Acme for one year under its primary direction and control, they must be treated as Acme's employees when determining whether Acme's retirement plan is tax qualified. Thus, Acme now has 40 workers for purposes of the rules. If it provides benefit to only 15, the rules won't be satisfied and the plan will lose its tax qualification. Acme won't be able to deduct its contributions to the plan and covered employees will have to pay tax on those contributions when Acme makes them.

There is one important exception to this rule. A leased employee will not be treated as your employee if:

- leased employees make up no more than 20% of the firm's nonhigh-compensation workforce
- the leased employees are covered under a tax-qualified pension plan provided by the leasing company, and
- the leasing organization's pension plan meets certain contribution, participation, and vesting requirements.

This is a very complex area of the law. If you have a retirement and/or profit-sharing plan for your employees and plan to lease workers from a leasing organization (or have already done so), get some help from your retirement plan administrator, pension plan consultant, or attorney or CPA specializing in the field.

RESOURCE

Get free information from the IRS. For more information on the tax qualification requirements for retirement plans, refer to IRS Publication 560, *Retirement Plans for Small Business*. You can download it from the IRS website at www.irs.gov.

Dealing With Established Leasing Companies

It's very important that you deal with a reputable leasing firm that is an established business with its own offices and management. Such a firm is more likely to make all required payroll tax and insurance payments. Remember, if the leasing firm doesn't pay these items, you may have to pay them.

Don't take a leasing firm's word that it's paying payroll taxes and insurance premiums. Require any leasing firm to provide you with proof that it is withholding and paying federal and state payroll taxes and paying for workers' compensation insurance for your leased employees.

Using Written Agreements

Sign a written lease agreement with the leasing company. Among other things, such an agreement should provide that:

- The leased workers are the leasing company's employees.
- The leasing company is responsible for paying the leased workers' wages and withholding and paying all state and federal payroll taxes, including unemployment compensation.
- The leasing company will provide the leased workers with workers' compensation insurance.
- The leasing company will indemnify you—that is, repay you—for all losses you might suffer as a result of its failure to comply with any legal requirements, including the costs of defending against charges of alleged violations. For example, if the leasing company fails to withhold and pay employment taxes, this provision will require it to pay any IRS assessments and penalties and your legal fees incurred in defending yourself against the IRS.

The agreement should also provide that the leasing company has the sole authority to hire, fire, schedule, supervise, and discipline the leased workers. (See above.)

Most leasing companies have their own lease agreements. It's wise to have an attorney review any leasing agreement before you sign it.

Procedures for Working With Independent Contractors

Before Hiring an IC 262

    Interviewing Prospective ICs 262

    Required Documentation 263

    Determining Whether Workers Qualify as ICs 264

    Drafting and Signing an IC Agreement 265

    Backup Withholding 265

    Keeping Records 267

While the IC Works for You 267

    Work Habits to Avoid 267

    Reimbursing an Independent Contractor for Expenses 269

    Adequate Accounting for Travel and Entertainment Expenses 270

    No Adequate Accounting for Travel and Entertainment Expenses 270

After the IC's Services End 271

    IRS Form 1099 271

    IRS Forms 4669 and 4670 279

f you hire ICs, you should assume that, sooner or later, the IRS and other government agencies may question the status of workers you've classified as ICs. Long before you're audited, you should have all the information and documentation you need to prove that a worker is an IC. Don't wait until you're audited to start thinking about how to prove a worker is an IC—by then it may be too late.

Before Hiring an IC

Someone in your company should be in charge of:

- interviewing prospective ICs
- determining whether applicants qualify as ICs
- authorizing workers to be hired as ICs, and
- preparing a file containing the information and documentation you'll need to prove that a worker is an IC if you're audited.

Whoever you choose should be fully trained regarding the laws and rules used to determine a worker's status. If you're running a one-person business, this person is you.

Interviewing Prospective ICs

All prospective ICs should fill out the Independent Contractor Questionnaire contained in Appendix B. (You can download the questionnaire on the Nolo website; the link is included in Appendix A.) Do not have an IC fill out an employment application; this makes the worker look like an employee.

Your contract administrator should review the questionnaire with the IC during an initial interview.

Questions You Shouldn't Ask

Federal and state laws bar employers from asking certain types of questions in interviews or on employment applications. Even though these laws don't apply to ICs, it's a good idea to follow them. For example, the Americans with Disabilities Act prohibits most preemployment questions about an applicant's disability. In addition, the Civil Rights Act forbids you to ask about an applicant's race, marital status, gender, birthplace, or national origin. There are also restrictions concerning questions about an applicant's age, arrest record, citizenship, and affiliations.

Required Documentation

Ask any worker you plan to hire as an IC to provide as many of the following documents as is possible and reasonable under the circumstances. Obviously, you'll want more documentation from ICs being hired to perform substantial projects than for those who do small ones. The relevant documents include:

- copies of the IC's business license (if required) and any professional licenses the IC has, such as a contractor's license
- certificates showing that the IC has insurance, including general liability insurance and workers' compensation insurance if the IC has employees
- the IC's business cards and stationery
- the URL for the IC's website, if any, and a screenshot of the first page
- copies of any advertising the IC has done, such as a yellow pages listing
- a copy of the IC's white pages business phone listing, if there is one
- if the IC is operating under an assumed name, a copy of the fictitious business name statement
- the IC's invoice form used for billing
- a copy of any office lease
- a photograph of the IC's office or workplace

- the IC's unemployment insurance number issued by the state unemployment insurance agency (only ICs with employees will have these)
- copies of 1099 forms issued to the IC by other companies for which the IC has worked
- the names and salaries of all assistants that the IC will use on the job
- the names and salaries of all assistants the IC has used on previous jobs for the past two years and proof that the IC has paid them, such as copies of canceled checks or copies of payroll tax forms
- a list of all the equipment and materials the IC will use in performing the services and how much they cost (proof that the IC has paid for the equipment, such as copies of canceled checks, is very helpful), and
- the names and addresses of other clients or customers for whom the IC has performed services during the previous two years (but don't ask for the identities of any clients the IC is required to keep confidential).

Determining Whether Workers Qualify as ICs

You must examine the answers the worker provided on the questionnaire, the documentation, and the task the IC is being hired to perform to see whether the worker can qualify as an IC.

Because there is no single definition of an IC, this can be a tough call. Some firms will hire any worker as an IC if he or she is incorporated and works no longer than six months for the firm. While both these factors are very helpful, they are no guarantee that the worker will not be reclassified as an employee by government auditors. You must weigh all the facts and circumstances on a case-by-case basis.

Use the Worker Classification Checklist in Appendix B to determine how to classify each worker. (You can download the checklist on the Nolo website; the link is included in Appendix A.) After you complete it, place it in your IC file.

Check on Safe Harbor Protections

If you can obtain safe harbor protection, you may treat the worker as an IC for employment tax purposes regardless of whether he or she qualifies as such under the normal IRS tests. Safe harbor protection is available only if you have consistently treated all workers performing similar services as ICs.

One very important question to ask, therefore, is whether your company has ever used employees to perform services similar to those the IC will be asked to do. If you have, you can forget about safe harbor protection—and you will likely have a harder task dealing with the IRS if you're audited. (See Chapter 3 for more information about safe harbor protection.) You're much better off keeping the work your employees and ICs do separate.

Drafting and Signing an IC Agreement

If you determine that the worker qualifies as an IC, complete and sign an independent contractor agreement before the IC starts work. (See Chapter 12 for guidance on creating an independent contractor agreement.) The IC may have his or her own agreement. If so, ask for a copy and use it as your starting point in drafting the agreement. This will show that the agreement is a real negotiated contract, not a standard form you forced the worker to sign. Pay particular attention to whether the IC's agreement contains any provisions that should be deleted or amended, or whether new provisions should be added.

Backup Withholding

Sometimes, you may have to withhold money from the IC and give it to the IRS. This is called backup withholding, and you can avoid it if you take the steps described below. If you do find yourself in the position of having to do backup withholding, follow the procedures described below.

How to Avoid Backup Withholding

It's very easy to avoid backup withholding. Have the IC fill out and sign IRS Form W-9, *Request for Taxpayer Identification Number and Certification*, then keep it in your IC file. You don't have to file the W-9 with the IRS. This simple form merely requires the IC to list his or her name, address, and taxpayer ID number. Corporations, partnerships, and sole proprietors with employees must have a federal employer identification number (EIN), which they obtain from the IRS. For sole proprietors without employees, the taxpayer ID number is either the IC's Social Security number or an EIN if the IC has obtained one.

A sole proprietor must always furnish his or her individual name on Form W-9, Line 1, regardless of whether the proprietor uses a Social Security number or an EIN. A sole proprietor may also provide a business name or "doing business as" (dba) name on Line 2, but must list the proprietor's individual name first.

If the IC doesn't already have an EIN or chooses to apply for one before filling out Form W-9, you don't have to backup withhold for 60 days after the date of the EIN application.

It's a good idea to check whether the taxpayer identification number (TIN) a worker provides you is correct before you submit a 1099-MISC for that worker. Use the IRS's online TIN matching service, which matches the name and number the worker gives you against IRS records. To use the service, you must enroll in the IRS e-services program and have filed at least one 1099-MISC during the previous two years. For more information, refer to IRS Publication 2108A, *On-Line Taxpayer Identification Number (TIN) Matching Program*, available at www.irs.gov.

Backup Withholding Procedure

If you are unable to obtain an IC's taxpayer ID number or the IRS informs you that the number the IC gave you is incorrect, you'll have to do backup withholding. Backup withholding must begin after you pay an IC $600 or more during the year. You need not backup withhold on payments totaling less than $600.

For this procedure, you withhold 28% of the IC's compensation and deposit it every quarter with your bank or other payroll tax depository. You must make these deposits separately from the payroll tax deposits you make for employees.

Report the amounts withheld on IRS Form 945, *Annual Return of Withheld Federal Income Tax*. This is an annual return you must file by January 31 of the following year. (See the instructions to Form 945 for details.)

Keeping Records

Create a file for each IC you hire. Keep these files separate from the personnel files you use for employees. Each file should contain:

- the signed final IC agreement and copies of any interim drafts
- the IRS W-9 form signed by the IC containing the IC's taxpayer identification number
- all the documentation provided by the IC, such as proof of insurance, business cards and stationery, copies of advertisements, professional licenses, and copies of articles of incorporation
- all the invoices the IC submits for billing purposes, and
- copies of all 1099 forms you file reporting your payments to the IC. (See below.)

Keep your IC files for at least six years.

While the IC Works for You

Treat the worker as an IC while he or she works for you—much the way you would the accountant who does your company's taxes or the lawyer who handles your legal work.

Work Habits to Avoid

There are a number of work habits you must avoid:

- Don't supervise the IC or his or her assistants. The IC should perform services without your direction. Your control should be limited to accepting or rejecting the final results.

- Don't let the IC work at your offices unless the nature of the services absolutely requires it—for example, if a computer consultant must work on your computers or a carpet installer is hired to lay carpet in your office.
- Don't give the IC employee handbooks or company policy manuals. If you need to provide ICs with orientation materials or suggestions, copies of governmental rules and regulations, or similar items, put them all in a separate folder titled "Orientation Materials for Independent Contractors" or "Suggestions for Independent Contractors."
- Don't establish the IC's working hours.
- Avoid giving ICs so much work or such short deadlines that they have to work full time for you. It's best for ICs to work for others at the same time they work for you.
- Don't provide ongoing instructions or training. If the IC needs special training, he or she should not obtain it in-house and should pay for it himself or herself.
- Don't provide the IC with equipment or materials unless absolutely necessary.
- Don't give an IC business cards or stationery to use that have your company name on them.
- Don't list or refer to ICs on your company website in ways that could be viewed as implying an employment relationship—for example, as "staff," a "member of the team," or "our people."
- Don't issue company email addresses or voicemail boxes to ICs.
- Don't give an IC a title within your company.
- Don't pay the IC's travel or other business expenses. Pay the IC enough to cover these expenses out of his or her own pocket.
- Don't give an IC benefits such as health insurance. Pay ICs enough to provide their own benefits.
- Don't require formal written reports. An occasional phone call inquiring into the work's progress is acceptable. But requiring regular written status reports indicates the worker is an employee.
- Don't invite an IC to employee meetings or functions.

- Don't refer to an IC as an employee, or to your company as the IC's employer, either verbally or in writing.
- Don't pay ICs on a weekly, biweekly, or monthly basis as you pay employees. Rather, require all ICs to submit invoices to be paid for their work. Pay the invoices at the same time you pay other outside vendors.
- Obey the terms of your IC agreement. Among other things, this means that you can't fire the IC. You can only terminate the IC's contract according to its terms—for example, if the IC's services fail to satisfy the contract specifications.
- Don't give the IC new projects after the original project is completed without signing a new IC agreement.

Reimbursing an Independent Contractor for Expenses

Independent contractors often incur expenses while performing services for their clients—for example, for travel, photocopying, phone calls, or materials. Many ICs want their clients to reimburse them separately for such expenses. The best practice is not to do this. It's better to pay ICs enough so they can cover their own expenses, rather than paying them less and having them bill you separately for expenses. ICs who pay their own expenses are less likely to be viewed as your employees by the IRS or other government agencies.

However, it's customary in some businesses and professions for the client to reimburse the IC for expenses. For example, a lawyer who handles a business lawsuit will usually seek reimbursement for expenses such as photocopying, court reporters, and travel. In this situation, you may reimburse the lawyer for these expenses.

When you reimburse an IC for a business-related expense, you get the deduction for the expense, and the IC does not. As long as the IC follows the adequate accounting rules discussed below, you do not include the amount of the reimbursement on the 1099 form you file with the IRS, reporting how much you paid the IC. The reimbursement is not considered income for the IC. Make sure to require ICs to document expenses with receipts, and save them in case the IRS questions the payments.

Adequate Accounting for Travel and Entertainment Expenses

The amount of deductible expenses you can list on your return depends on whether the IC provides you with an "adequate accounting." To make an adequate accounting of travel and entertainment expenses, an IC must comply with all the record-keeping rules applicable to business owners and employees. The IRS is particularly suspicious of travel, meal, and entertainment expenses, so there are special documentation requirements for these. You are not required to save the IC's expense records except for records for entertainment expenses.

You may deduct the IC's travel, entertainment, and meal expenses that you have reimbursed as your own business expenses. Remember that meal and entertainment expenses are only 50% deductible.

> EXAMPLE: Tim hires Mary, a self-employed marketing consultant, to help him increase his business's sales. In the course of her work, Mary incurs $1,000 in meal and entertainment expenses while meeting potential customers. She makes an adequate accounting of these expenses and Tim reimburses her the $1,000. Tim may deduct 50% of the $1,000 as a meal-and-entertainment expense for his business; Mary gets no deduction. When Tim fills out the 1099 form reporting to the IRS how much he paid Mary, he does not include the $1,000. Tim should save all the documentation Mary gave him to prove her entertainment expenses.

No Adequate Accounting for Travel and Entertainment Expenses

If an IC doesn't properly document travel, meal, or entertainment expenses, you do not have to keep records of these items. You may reimburse the IC for the expenses and deduct the full amount as IC payments, provided they are ordinary, necessary, and reasonable in amount. You are deducting these expenses as IC compensation, not as travel, meal, or entertainment expenses, so the 50% limit on deducting meal and entertainment costs does not apply. You must include the amount of the reimbursement as income paid to the IC on the IC's 1099 form. Clearly, you are better off if the IC doesn't adequately account for travel, meal, and entertainment expenses—but the IC is worse off because the IC must pay tax on the reimbursements.

> **EXAMPLE:** Assume that Mary (from the example above) incurs $1,000 in meal and entertainment expenses but fails to adequately account for these expenses to Tim. Tim reimburses her the $1,000 anyway. Tim may deduct the full $1,000 as a payment to Mary for her IC services. Tim must include the $1,000 as a payment to Mary when he fills out her 1099 form, and Mary must pay tax on the money. She may deduct the expenses as business expenses on her own tax return, but her meal and entertainment expenses will be subject to the 50% limit. Moreover, she'll need to have adequate documentation to back up the deductions if she is audited by the IRS.

After the IC's Services End

A hiring firm's work is never done. Even after an IC's services end, there is paperwork to complete. Failing to complete it could lead to severe penalties if you're audited.

IRS Form 1099

The single most important thing to do after an unincorporated IC's services end is to provide the worker and IRS with an IRS Form 1099-MISC reporting the compensation you paid the worker. Your failure to do so will result in severe penalties if the IRS later audits you and determines you misclassified the worker:

- You'll be required to pay the IRS twice as much in penalties for the misclassification—that is, if you had filed the 1099 forms you would owe half as much.

- You'll lose the right to safe harbor protection for any payments not reported on Form 1099. This means you'll lose one of your most valuable legal rights in defending against the IRS.
- The IRS may impose a $100 fine for each Form 1099 you failed to file.

When Form 1099 Must Be Filed

The basic rule is that you must file a Form 1099 whenever you pay an unincorporated IC—that is, an IC who is a sole proprietor or member of a partnership or limited liability company—$600 or more in a year for work done in the course of your trade or business.

EXAMPLE: The Acme Widget Company hires Thomas to install a new computer system. Acme classifies Thomas as an IC and pays him $2,000. Thomas is a sole proprietor. Because Acme paid him more than $599, it must file a Form 1099 with the IRS reporting the payment.

In calculating whether the payments made to an IC total $600 or more during a year, you must include payments for parts or materials used by the IC in performing the services.

EXAMPLE: The Old Reliable Insurance Company pays $1,000 to an unincorporated auto repair shop to repair one of its insured's cars. The repair contract states that $300 is for labor and $700 is for parts. Old Reliable must report the entire $1,000 on Form 1099 because the parts were used by the repair shop to perform the car repair services.

Services in the Course of Your Trade or Business

You need to file a Form 1099 only if an IC performs services that are in the course of your trade or business. A trade or business is an activity carried on for gain or profit. You don't have to file a Form 1099 for payments for nonbusiness-related services. This includes payments you make to ICs for personal or household service—for example, payments to babysitters, gardeners, and housekeepers. Running your home is not a profit-making activity.

EXAMPLE 1: Eddy owns several homes that he rents out to tenants. He is in the business of renting houses. Eddy pays Linda, who operates a painting business as a sole proprietor, $1,750 to paint one of his rental houses. Eddy must report the $1,750 payment to Linda on Form 1099.

EXAMPLE 2: Eddy pays Linda $1,750 to paint his own home. He lives in this home with his wife and family and it is not a part of his home rental business. Eddy need not report this payment on Form 1099 because this work was not done in the course of his business.

Although nonprofit organizations are not engaged in a trade or business, they are still required to report payments made to ICs. Payments to ICs by federal, state, or local government agencies must also be reported.

EXAMPLE: Leslie, owner of an unincorporated home repair business, is paid $1,000 for repair work on a church. Even though the church is a tax-exempt nonprofit organization and is not engaged in business, it must report the $1,000 payment to Leslie on Form 1099.

Payments Exempt From Form 1099 Filing Requirement

You don't need to report on Form 1099 payments solely for merchandise or inventory. This includes raw materials and supplies that will become a part of merchandise you intend to sell.

EXAMPLE: The Acme Widget Company pays $5,000 to purchase 100 used widgets from Joe's Widgets, a sole proprietorship owned by Joe. Acme intends to fix up and resell the widgets. The payment to Joe need not be reported on Form 1099 because Acme is purchasing merchandise from Joe, not IC services.

In addition, you need not file a Form 1099 when you pay for:
- freight, storage, and similar items
- telephone, telegraph, and similar services, or
- rent to real estate agents.

You need not file 1099 forms for wages paid to your employees or for payments for their traveling or business expenses. These amounts are reported on Form W-2.

No Form 1099 for Corporations

You need not file a Form 1099 for payments made to an incorporated IC. This is one of the main advantages of hiring incorporated ICs, because the IRS uses Form 1099 as an audit lead.

> **EXAMPLE:** The Acme Sandblasting Company pays $5,000 to Yvonne, a CPA, to perform accounting services. Yvonne has formed her own one-person corporation called Yvonne's Accounting Services, Inc. Acme pays the corporation, not Yvonne personally. Because Acme is paying a corporation, it need not report the payment on Form 1099.

However, payments to certain types of corporations must be reported:
- Payments to medical corporations must be reported on Form 1099 (including payments made in the course of business to incorporated veterinarians).
- Payments made to attorneys who have incorporated their practices must be reported on Form 1099.
- All federal executive agencies must prepare a Form 1099 for services performed by corporations if the corporation is paid $600 or more during the calendar year.

CAUTION

Get the corporation's full name and EIN. Make sure you have an IC corporation's full legal name and federal employer identification number. Without this information, you may not be able to prove to the IRS that the IC was incorporated. An easy way to do this is to have the IC fill out IRS Form W-9, *Request for Taxpayer Identification Number and Certification*, and keep it in your files. This simple form merely requires the IC to provide his or her corporation's name, address, and federal EIN.

Payments to Lawyers

If you pay $600 or more in one year to an attorney in the course of your trade or business—for example, to represent your company in a lawsuit or to draft contracts for your company—you must report the payment(s) on IRS Form 1099. This is true whether the attorney works as a sole proprietor, an LLC member, a member of a partnership, or an employee of a corporation. You do not have to report payments to attorneys for personal matters—for example, to handle your divorce.

There might be times when you will pay money to an attorney who does not represent your company—for example, an attorney who represents a plaintiff who sues your company. If you give an attorney who does not represent you a lump sum of money, but don't know how much of that money will go to the attorney or how much will go to the attorney's client, you must report the entire amount in Box 14 on Form 1099. For example, if you pay $100,000 to an attorney to settle a case but don't know how much of the money will go to the attorney, you must report the entire amount in Box 14 of Form 1099. If, on the other hand, you know that the lawyer's fee is $34,000, you would report only that amount in Box 7 of Form 1099.

Filing Procedures

You must file one 1099 form for each IC to whom you paid $600 or more during the year. You must obtain original paper 1099 forms from the IRS or use electronic forms; you cannot photocopy this form. Each 1099 form contains three parts and can be used for three different workers. All your 1099s must be submitted together along with one copy of IRS Form 1096, which is a transmittal form—the IRS equivalent of a cover letter. (See below for a sample.) You must obtain an original paper Form 1096 from the IRS or use an electronic form; you cannot submit a photocopy.

Filling out the 1099 form is easy. Follow this step-by-step approach:

- List your name and address in the first box titled "Payer's name."
- Enter your taxpayer identification number in the box entitled "Payer's federal identification number."
- The IC is called the "Recipient" on this form, meaning the person who received payment from you. You must provide the IC's taxpayer identification number, name, and address in the boxes indicated. For sole proprietors, you must list the individual's name first; you can also list a different business name. You may not enter only a business name for a sole proprietor.
- You must enter the amount of your payments to the IC in Box 7, entitled "Nonemployee compensation." Be sure to fill in the right box or the IRS will deem the 1099 form invalid.
- Finally, if you've done backup withholding for an IC who has not provided you with a taxpayer ID number, enter the amount withheld in Box 4.

The 1099-MISC form contains five copies. These must be filed as follows:

- Copy A, the top copy, must be filed with the IRS no later than February 28 of the year after you paid the IC. Don't cut or separate this page, even though it has spaces for three workers.
- Copy 1 must be filed with your state taxing authority if your state has an income tax. The filing deadline is probably February 28, but check with your state tax department to make sure. Your state may have a specific transmittal form you must obtain.
- You must give Copy B and Copy 2 to the worker no later than January 31 of the year after payment was made.
- Copy C is for your files.

All the IRS copies of each 1099 form are filed together with Form 1096, a simple transmittal form. You must add up all the payments reported on all the 1099s and list the total in the box indicated on Form 1096. File the forms with the IRS Service Center listed on the reverse of Form 1096.

If you wish, you may file your 1099s electronically instead of by mail. You must first get permission from the IRS to do this by filing IRS Form 4419, *Application for Filing Information Returns Electronically.* However, if you must file 250 or more 1099s, you *must* file them electronically. If you file electronically, the deadline for filing 1099s is March 31. For more information, you can visit the IRS website at www.irs.gov or call the IRS at 866-455-7438.

You may send 1099s to independent contractors via email, but only if the ICs agree. Otherwise, you must deliver the form in person or by postal mail.

Filing 1099s Online

The IRS doesn't have an electronic 1099-MISC on its website that you can fill out and e-file. Instead, you must use specialized software for this purpose.

There are small-business-accounting software and payroll applications through which you can file your 1099s electronically (or by postal mail). There are also several websites specializing in such filings.

CAUTION

When in doubt, file a 1099. If you're not sure whether a 1099 form must be filed for a worker, go ahead and file one anyway. You lose nothing by doing so and will save yourself the severe consequences of not filing if you were in fact required to do so.

Getting More Time to File 1099s

You can obtain a 30-day extension of the time to file a 1099 form by filing IRS Form 8809, *Application for Extension of Time To File Information Returns.* The form must be filed with the IRS by February 28 (or March 31, if filing electronically). The extension is not granted automatically; you must explain why you need it. The IRS will send you a letter of explanation approving or denying your request. See the instructions for Form 8809 for more information.

IRS Forms 4669 and 4670

If the IRS determines that you intentionally misclassified a worker as an IC, it will impose a 20% income tax assessment, which means you will be required to pay the IRS an amount equal to 20% of all the payments you made to the worker.

However, this assessment will be reduced or eliminated if you can prove the worker reported and paid income taxes on the payments. Such a reduction is called an offset or abatement.

To obtain this abatement, you must file IRS Form 4669, *Statement of Payments Received*. This form states how much tax the worker paid on the wages in question. The worker must sign the form under penalty of perjury. You must file a Form 4669 for each worker involved along with Form 4670, *Request for Relief from Payment of Income Tax Withholding*, which is used to summarize and transmit the Form 4669.

Unfortunately, by the time your company is audited—typically one to three years after you hired the worker—it may be impossible for you to locate the worker or persuade him or her to complete and sign a Form 4669. This means you'll be unable to get the abatement.

A better approach is to be proactive and ask workers to sign a Form 4669 as soon as their services end. Retain the signed 4669s in your IC files. This way you'll be able to obtain an abatement if the worst happens: You're audited by the IRS and it claims you have intentionally misclassified the worker.

At the latest, by April 15 of the year following the year an IC performed services for you, send him or her a blank Form 4669 to fill out and sign. Your IC agreement should contain a provision obligating the worker to provide the information required in the form. (See Chapter 12 for guidance on creating IC agreements.)

State New-Hire Reporting Requirements
for Independent Contractors

Several states require businesses that hire independent contractors to file a report with a state agency providing the contractor's contact information and how much the worker is paid. The purpose of these requirements is to help the state enforce child support orders issued against independent contractors.

The following states impose reporting requirements for those who hire independent contractors: Alabama, California (if the independent contractor is paid over $600 per year), Connecticut (if the independent contractor is paid over $5,000 per year), Iowa, Massachusetts (if the independent contractor is paid over $600 per year), Nebraska, New Hampshire, New Jersey, Ohio (if the independent contractor is paid $2,500 or more per year), and West Virginia. Find the contact information for your state agency at: www.acf.hhs.gov/sites/default/files/programs/css/state_new_hire_reporting_contacts_and_program_information.pdf.

Independent Contractor Agreements

Using Written Agreements 283

   Oral Agreements 283

   Establishing IC Status 284

Drafting Agreements 285

   Standard Form Agreements 285

   The Drafting Process 286

   Putting the Agreement Together 288

Essential Provisions 292

   Term of Agreement 294

   Services to Be Performed 294

   Payment 295

   Terms of Payment 297

   Expenses 299

   Independent Contractor Status 300

   Business Permits, Certificates, and Licenses 302

   State and Federal Taxes 302

   Fringe Benefits 303

   Workers' Compensation 304

   Unemployment Compensation 305

   Insurance 306

   Terminating the Agreement 307

   Exclusive Agreement 308

   Severability 308

   Applicable Law and Forum 309

   Notices 309

   No Partnership 310

Assignment..311

Resolving Disputes..313

Signatures..317

Optional Provisions...318

Modifying the Agreement..318

Work at Your Premises..320

Indemnification...320

Intellectual Property Ownership..321

Confidentiality 324

Nonsolicitation...325

Sample IC Agreement...325

ThisT chapter explains why you should use written agreements with independent contractors, describes what such agreements should contain, and provides sample language for you to use.

You can draft contracts using the suggested text offered here, choosing the clauses that apply to your situation. You can then use your final draft as a starting point of negotiations with the IC and tailor your final agreement to both of your needs.

> **FORM**
>
> **You can download copies of all the agreements in this chapter on the Nolo website;** the link is included in Appendix A.

> **RESOURCE**
>
> **For additional sample IC agreements for workers in over a dozen occupations,** refer to *Consultant & Independent Contractor Agreements*, by Stephen Fishman (Nolo).

Using Written Agreements

You should sign a written agreement with every IC you hire, before he or she starts work. An IC agreement serves two main purposes:

- It avoids later disputes by providing a written description of the services the IC is supposed to perform and how much the IC will be paid.
- It describes the relationship between you and the IC to help make clear that the IC is not your employee.

Oral Agreements

Courts are crowded with lawsuits filed by people who entered into oral agreements with one another and later disagreed over what was said. Costly misunderstandings can develop if an IC performs services for you without a document clearly stating what he or she is supposed to do and what will happen if it isn't done. Such misunderstandings may be innocent; you and the IC may have simply misinterpreted one another. Or they may be purposeful; without a written record to contradict him or her, an IC can claim that you orally agreed to anything.

Consider a good written IC agreement your legal lifeline. If disputes develop, the agreement will provide ways to solve them. If you and the IC end up in court, a written agreement will establish your legal duties to one another.

Some Agreements Must Be in Writing

Some types of agreements must be in writing to be legally enforceable. Each state has a law, usually called the statute of frauds, listing the types of contracts that must be in writing to be valid. These lists often include:

- Any contract that cannot possibly be performed in less than one year.

  EXAMPLE: John agrees to perform consulting services for Acme Corp. for the next two years for $2,000 per month. The agreement cannot be performed in less than one year, so it must be in writing to be legally enforceable.

- Contracts for the sale of goods—that is, tangible personal property such as a computer or car—worth $500 or more.
- A promise to pay someone else's debt. For example, if the president of a corporation personally promises to pay for the services you sell the corporation, the guarantee must be in writing to be legally enforceable.
- Contracts involving the sale of real estate, or real estate leases lasting more than one year.

Federal copyright law applicable in every state also requires that any transfer of copyright ownership be in writing to be valid.

Establishing IC Status

If your business is audited, an auditor will ask to see your agreements with all workers you've classified as independent contractors. A well-drafted agreement that indicates that the IC is in business for himself or herself will prove that the worker is an independent contractor. The document will help demonstrate that you and the worker intended to create an independent contractor relationship. And because written agreements with ICs have become a routine fact of business life, you will immediately look suspect if you don't have one.

However, a written IC agreement is not a magic legal bullet. It won't make a worker an IC all by itself. What really counts is how you treat a worker. If the agreement is accurate when it says the worker is in business, that will be helpful. But it won't help a bit if you treat the worker like an employee.

> **EXAMPLE:** AcmeSoft, Inc., hires Pat to perform computer programming. It requires her to sign a document called an Independent Contractor Agreement. The agreement states that Pat is an IC and that AcmeSoft will exercise no control over Pat on the job. However, in reality, Pat is treated just like an AcmeSoft employee: Her work is closely supervised, she is paid biweekly, she has set hours of work, and she works only for AcmeSoft. Pat is an AcmeSoft employee, despite what the agreement says.

Drafting Agreements

You don't need to hire a lawyer to draft an independent contractor agreement. This chapter gives you guidance and suggested language to use as a starting point in fashioning an agreement that will meet your needs.

Standard Form Agreements

There are various standard form IC agreements you can obtain online and from other sources. These forms contain standard one-size-fits-all legalese and are not tailored for any particular occupation.

IRS and state auditors are well aware that businesses often have workers sign such generic IC agreements before they start work. The workers may not even bother to read the agreements and certainly won't change them to reflect the real work situation.

The more an IC agreement is custom-tailored to the true relationship between the worker and the hiring business, the more helpful it will be.

The Drafting Process

Either you or the IC should begin the process of drafting an agreement by offering the other person proposed language to include. Then both of you can negotiate and make changes until you agree on a final version.

Using the IC's Agreement

Many ICs have their own agreements they've used in the past. If so, it's often wise to use that agreement as the starting point. This avoids giving government auditors the impression that the agreement is simply a company-drafted standard form the worker was forced to sign.

Read the IC's agreement carefully, because it may contain provisions unduly favorable to the IC and harmful to you. Make sure it contains all the necessary provisions discussed in this chapter. You may also wish to add some provisions of your own or other optional provisions, as discussed below.

Keep copies of the IC's original agreement and all of the drafts and changes. These will help show auditors that the agreement was negotiated, not a contract you imposed on the IC.

Using Your Own Agreement

If the IC does not have an agreement you can use, you'll have to provide one of your own. You can use the general IC agreement provided here and adapt it for almost any kind of work.

Again, feel free to add provisions of your own or other options discussed below.

Electronic Contracts

Traditionally, contracts have been in the form of paper copies, hand-signed in ink. However, contracts in this form aren't legally required—electronic agreements without paper or other hard copies are perfectly legal. Using electronic agreements can be much faster and easier than dealing with paper. For example, you may draft a contract on your computer and email a digital copy to an IC. Once the IC emails it back with a digital signature indicating acceptance, there is no need to deal with the post office or a courier service to deliver a copy.

The validity of electronic agreements is secured by federal Electronic Signatures in Global and National Commerce Act (the E-Sign Act). In addition, all states have adopted the Uniform Electronic Transactions Act (UETA), which establishes the legal validity of electronic signatures and contracts in a way similar to the federal law. This ensures that electronic contracts and electronic or digital signatures will be valid in all states, regardless of where the parties live or where they execute the contract.

Careful—Contracts by Email

You can inadvertently create a legally binding contract through a simple exchange of emails, even if neither party prints and physically signs anything. A binding contract can come into existence whenever when one party promises to provide goods or services and the other promises to pay for them.

So, be careful. If you send an IC an email asking him or her to perform X services for Y money and the IC emails back, "OK," you may have created a legally binding contract. If you later decide that Y money is too much, you could be out of luck. To avoid this trap, say in your first email something like, "If this sounds good, let's draft an agreement to hammer out the details."

Signing Electronic Contracts

A traditional ink signature isn't possible on an electronic contract—you must use a digital or electronic signature. Electronic and digital signatures—which differ from each other—can be accomplished in a variety of ways.

Digital signatures are created via cryptographic "scrambling" technology to ensure that the e-signature really belongs to the person who was supposed to sign the contract. Various software packages—both free and commercially distributed—are available to make using digital signatures easy.

"Electronic signature" refers to any means of creating a signature other than using sophisticated cryptography. It includes, for example, using computer or mobile applications to upload a scanned version of the signer's signature, typing the signer's name into a signature area, or signing with a computer mouse or one's finger. Many applications are available for this purpose.

Putting the Agreement Together

Make sure your IC agreement is properly signed and put together. This is not difficult if you know what to do. Follow the tips offered below.

Signatures

It's best for both you and the IC to sign the agreement. Signatures should be in ink. You and the IC need not be together when you sign, and it isn't necessary for you to sign at the same time. There's no legal requirement that the signatures be located in any specific place in a business contract, but they are customarily placed at the end of the agreement; that helps show that you both have read and agree to the entire document.

It's very important that both you and the IC sign the agreement properly. Failure to do so can have drastic consequences. How to sign depends on the legal form of your businesses.

- **Sole proprietors.** If you and the IC are sole proprietors, you can simply sign your own names, because a sole proprietorship is not a separate legal entity. However, if a sole-proprietor IC uses a fictitious business name, it's better for him or her to sign on behalf of the business. This will help show the worker is an IC, not your employee.

    EXAMPLE: You hire Chris Craft, a sole-proprietor IC who runs a marketing research business. Instead of using his own name for the business, he calls it AAA Marketing Research. You should have him sign the IC agreement like this:

    > AAA Marketing Research
    >
    > By: *Chris Craft*
    >   Chris Craft

- **Partnerships.** If either you or the IC is in a partnership, a general partner should sign on the partnership's behalf. Only one partner needs to sign. The signature block for the partnership should state the partnership's name and the name and title of the person signing on the partnership's behalf. If a partner signs only his or her name, without mentioning the partnership, the partnership is not bound by the agreement.

EXAMPLE: You hire Chris, the general partner of a partnership called The Chris Partnership, to perform marketing research. He should sign the contract on the partnership's behalf like this:

The Chris Partnership
A California Partnership

By: *Chris Craft*
     Chris Craft, General Partner

If an IC is a partnership and a person who is not a partner signs the agreement, the signature should be accompanied by a partnership resolution stating that the person signing the agreement has the authority to do so. The partnership resolution is a document signed by one or more of the general partners stating that the person named has the authority to sign contracts on the partnership's behalf.

- **Limited liability companies.** The owners of a limited liability company (LLC) are called "members." Members may hire others (called "managers") to run their LLC business for them. If you hire an IC who has formed an LLC, the agreement should be signed by a member or manager on the LLC's behalf.

  EXAMPLE: You hire Roger, an efficiency expert, to help streamline your business operations. Roger has formed an LLC to run his business called Efficiency Expertise, LLC. The signature block should appear in the contract like this:

Efficiency Expertise, LLC
An Illinois Limited Liability Company

By: *Roger Bacon*
     Roger Bacon, Member

- **Corporations.** If either you are or the IC is a corporation, the agreement must be signed by someone who has authority to sign contracts on the corporation's behalf. The corporation's president or chief executive officer (CEO) is presumed to have this authority.

  If someone other than the president of an incorporated IC signs—for example, the vice president, the treasurer, or another corporate officer—ask to see a board of directors' resolution or corporate bylaws authorizing him or her to sign. If the person signing doesn't have authority, the corporation won't be legally bound by the contract.

  If you sign personally instead of on your corporation's behalf, you'll be personally liable for the contract.

  The signature block for a corporation should state the name of the corporation and indicate the name and title of the person signing on the corporation's behalf.

  > **EXAMPLE:** You hire Chris Craft, an IC marketing consultant, to perform marketing research. Chris is president of his own corporation, called Chris Marketing, Inc. Chris should sign the IC agreement on behalf of his corporation. The signature block should appear in the contract like this:

  > Chris Marketing, Inc.
  > A California Corporation
  >
  > By: _Chris Craft_ _____
  >   Chris Craft, President

Dates

A contract must have a date. This can be in the first paragraph, or you can simply put a date line next to the place where each person signs—for example:

> Date: _____, 20___ .

The parties need not sign the contract the same day that it becomes effective or that performance must begin. Either of the latter can be a later date that's provided in the contract. In addition, you and the IC don't have to sign on the same day. Indeed, you can sign weeks apart.

Attachments or Exhibits

An easy way to keep the main body of an IC agreement as short as possible is to use attachments, also called exhibits. You can use them to list lengthy details such as performance specifications. This makes the main body of the agreement easier to read.

If you have more than one attachment or exhibit, they should be numbered or lettered—for example, Attachment 1 or Exhibit A. Be sure to mention that they're included as part of the contract in the main body of the agreement.

Changing the Contract

Sometimes it's necessary to make changes to a contract just before it's signed. If you use a computer to prepare the agreement, it's usually easy to make the changes and print out a new agreement.

However, it's not always necessary to prepare a new contract. Instead, the changes may be handwritten or typed onto all existing copies of the agreement. The changes should be initialed by those signing the agreement as close as possible to the place where the change is made. If both people who sign don't initial each change, questions might arise as to whether the change was part of the agreement.

Copies of the Contract

Prepare at least two originals of your agreement. Make sure that each contains all of the exhibits and attachments. Both you and the IC should sign both. Both you and the IC should keep one signed original of the agreement.

Faxing or Emailing Contracts

It is common for businesses and ICs to communicate by fax or email. They often send scanned drafts of their proposed agreements back and forth to each other. When a final agreement is reached, one signs a copy of the contract and faxes or emails it to the other, who signs it and faxes or emails it back.

A faxed or emailed signature by hand is probably legally sufficient if neither you nor the IC dispute that it is a representation of an original signature. However, if an IC claims that such a signature was forged, it could be difficult or impossible to prove that it's genuine because it is very easy to forge a faxed or emailed signature with modern computer technology. Forgery claims are rare, however, so this is usually not a problem. Even so, it's a good practice for you and the IC to follow up the fax or email with signed originals exchanged by postal mail or express delivery.

Essential Provisions

The provisions in this section should be included in most IC agreements. (See below to see how an entire agreement might look when assembled.)

This information can be used as:

- a checklist when you review an IC agreement provided by a worker to make sure nothing important has been left out, or
- a starting point to draft your own agreement.

These provisions may be all you need for a simple IC agreement. Or you may need to combine them with some of your own clauses or one or more of the optional clauses discussed below.

Title of agreement. Deceptively simple things such as what you call an IC agreement and how you refer to yourself and the IC can have a big impact. You need not have a title for an IC agreement, but if you want one, call it Independent Contractor Agreement or Consulting Agreement. Consulting Agreement sounds a little more high-toned than Independent Contractor Agreement and is often used when contracting with skilled professionals to provide services. For example, IC agreements with computer software experts are often called Consulting Agreements. Do not call your contract an Employment Agreement.

Suggested Language

Independent Contractor Agreement

FORM

You can download all of these clauses on the Nolo website; the link is included in Appendix A.

Names of IC and hiring firm. Do not refer to an IC as an employee or to yourself as an employer.

Initially, it's best to refer to the IC by his or her full name. If an IC is incorporated, use the corporate name, not the IC's own name—for example: "John Smith, Incorporated" instead of "John Smith." If the IC is unincorporated but does business under a fictitious business name, use that name. A fictitious business name or an assumed name is a name sole proprietors or partners use to identify their businesses. For example, if consultant Al Brodsky calls his one-person marketing research business ABC Marketing Research, use that name. This shows you're contracting with a business, not a single individual.

For the sake of brevity, it is usual to identify yourself and the IC by shorter names in the rest of the agreement. You can use an abbreviated version of the IC's full name—for example, ABC for ABC Marketing Research. Or you can refer to the IC simply as Contractor or Consultant.

Refer to yourself initially by your company name and subsequently by a short version of the name or as Client or Firm.

Also include the business addresses of the IC and yourself. If you have or the IC has more than one office or workplace, list the principal place of business—the main office or workplace.

Suggested Language

Names of IC and Hiring Firm

This Agreement is made between [*your company name*] (Client), with a principal place of business at [*your business address*] and [*IC's name*] (Contractor), with a principal place of business at [*IC's address*].

Term of Agreement

The term or duration of the agreement should be as short as possible. A good outside time limit is six months. A longer term makes the agreement look like an employment agreement, not an IC agreement. If the work is not completed at the end of six months, you can negotiate and sign a new agreement.

The date the agreement begins can be the date you sign it or another date after you sign.

Suggested Language

> **Term of Agreement**
>
> This Agreement will become effective on [*date*], and will end no later than the earlier of [*date*] and completion of Contractor's services under this Agreement.

Services to Be Performed

The agreement should describe in as much detail as possible what you expect the IC to do. You must word the description carefully to concentrate on the results you expect him or her to achieve. Don't tell the IC how to achieve the result; that would indicate that you have the right to control how the IC performs the work, which can create an employment relationship.

> **EXAMPLE:** Jack hires Jill to prepare an index for his multivolume history of ancient Sparta. Jack describes the results he expects Jill to achieve like this: "Contractor agrees to prepare an index of Client's *History of Sparta* of at least 100 single-spaced pages. Contractor will provide Client with a printout of the finished index and a version on compact disc in ASCII format."
>
> Jack should not tell Jill how to create the index. For example, Jack should not write: "Contractor will prepare an alphabetical three-level index of Client's *History of Sparta*. Contractor will first prepare 3" × 5" index cards listing every index entry beginning with Chapter One.

After each chapter is completed, Contractor will deliver the index cards to Client for Client's approval. When index cards have been created for all 50 chapters, Contractor will create a computer version of the index using Complex Software Version 7.6. Contractor will then print out and edit the index and deliver it to Client for approval."

It's perfectly okay for you to establish very detailed specifications for the IC's finished work product. But the specs should only describe the end results the IC must achieve, not how to obtain those results.

You can include the description in the main body of the agreement or in a separate attachment, if it's lengthy.

Suggested Language: Alternative A

Services to Be Performed

Contractor agrees to perform the following services: [*Briefly describe services you want performed by IC.*]

Suggested Language: Alternative B

Services to Be Performed

Contractor agrees to perform the services described in Exhibit A, which is attached to this Agreement.

Choose Alternative A if you include the explanation of services in the contract. Choose Alternative B if the explanation of the services is attached to the main contract.

Payment

Independent contractors are usually paid in one of two ways:
- a fixed fee, or
- by unit of time.

Fixed Fee

Paying an IC a fixed sum for the entire job, rather than an hourly or daily rate, strongly supports a finding of independent contractor status. If paid a fixed sum, the IC risks losing money if the project takes longer than expected, or may earn a substantial profit if the project is completed quickly. Having the opportunity to earn a profit or suffer a loss is a very strong indication of IC status.

Suggested Language: Alternative A

Payment

In consideration for the services to be performed by Contractor, Client agrees to pay Contractor $[*state amount*] according to the terms set out below.

Unit of Time

Paying a worker by the hour or another unit of time usually indicates that the worker is an employee because the worker has no real risk of loss. However, it's customary for ICs in some occupations—for example, lawyers and accountants—to charge by the hour. Government auditors would likely not challenge the IC status of workers in these occupations as long as they are in business for themselves.

Suggested Language: Alternative B

Payment

In consideration for the services to be performed by Contractor, Client agrees to pay Contractor at the rate of $[*state amount*] per [*hour, day, week, or other unit of time*] according to the terms of payment set out below.

If you pay by the hour, you may want to place a cap on the IC's total compensation. This may be a particularly good idea if you're unsure of the IC's reliability or efficiency.

Optional Addition

Unless otherwise agreed in writing, Client's maximum liability for all services performed during the term of this Agreement shall not exceed $[*state the top limit on what you will pay*].

Choose Alternative A to pay the IC a fixed fee or Alternative B to pay by the hour. Add the optional language if you wish to cap the IC's payments.

Terms of Payment

Because an IC is running an independent business, he or she should submit an invoice setting out the amount you have to pay.

Fixed Fee Agreements

The following provision requires you to pay the IC's fixed fee within a specific time after the work is completed. Add the time period for payment—for example, seven, 14, 30, 60, or 90 days.

Suggested Language: Alternative A

Terms of Payment

Upon completing Contractor's services under this Agreement, Contractor shall submit an invoice. Client shall pay Contractor the compensation described within [*list number—e.g., 7, 14, 30, 60, 90*] days after receiving Contractor's invoice.

Divided Payments

You can also opt to pay part of a fixed fee when the agreement is signed and the remainder when the work is finished.

The following provision allows you to pay a specific amount when the IC signs the agreement and then the rest when the work is finished. The amount of the up-front payment is subject to negotiation. It could be as little as 10% or less of the entire fixed fee.

Suggested Language: Alternative B

Terms of Payment

Contractor shall be paid $[*state amount*] upon signing this Agreement and the rest of the sum described above when Contractor completes services and submits an invoice.

Installment Payments

An IC may balk at accepting a fixed fee for a complex or long-term project due to difficulties in accurately estimating how long the job will take. One way to deal with this problem is to break the job into parts and pay the IC a fixed fee when each phase is completed. If, after one or two phases are completed, it looks like the fixed sum won't be enough to complete the entire project, you can always renegotiate the agreement.

This type of arrangement is far more supportive of an IC relationship than hourly payment, because the IC still has some risk of loss.

To do this, draw up a schedule of installment payments tying each payment to the IC's completion of specific services and attach it to the agreement. The main body of the agreement should simply refer to the attached payment schedule.

Suggested Language: Alternative C

Terms of Payment

Client shall pay Contractor according to the following schedule of payments

1. $[*State sum*] when an invoice is submitted and the following services are complete: [*Describe first stage of services*].
2. $[*State sum*] when an invoice is submitted and the following services are complete: [*Describe second stage of services*].
3. $[*State sum*] when an invoice is submitted and the following services are complete: [*Describe third stage of services*].

Invoices and Payments by Unit of Time

Even ICs who are paid by the hour or other unit of time should submit invoices. Do not automatically pay an IC weekly or biweekly the way you pay employees. It's best to pay ICs no more than once a month; this is how businesses are normally paid.

Suggested Language: Alternative D

Terms of Payment

Contractor shall submit an invoice to Client on the last day of each month for the work performed during that month. The invoice should include: an invoice number, the dates covered by the invoice, the hours expended, and a summary of the work performed. Client shall pay Contractor's fee within [*list number—e.g., 7, 14, 30, 60, 90*] after receiving the invoice.

Choose Alternative A to pay a fixed fee, Alternative B to pay in divided payments, Alternative C to pay in installments, or Alternative D if you will pay the IC according to the number of hours worked.

Expenses

An IC should usually not be reimbursed for expenses. Instead, compensate the IC well enough so that he or she can pay the expenses directly out of his or her own pocket.

However, it is customary to pay the expenses of certain types of ICs. For example, attorneys typically charge their clients separately for photocopying charges, deposition fees, and travel. Where there is an otherwise clear IC relationship and payment of expenses is customary in the IC's trade or business, you can probably get away with doing it.

Suggested Language

Expenses

Contractor shall be responsible for all expenses incurred while performing services under this Agreement. These include license fees, memberships, and dues; automobile and other travel expenses; meals and entertainment; insurance premiums; and all salary, expenses, and other compensation paid to employees or contract personnel Contractor hires to complete the work under this Agreement.

Independent Contractor Status

One of the most important functions of an independent contractor agreement is to help establish that the worker is an IC, not your employee. The key to doing this is to make clear that the IC has the right to control how the work will be performed.

You will need to emphasize the factors the IRS and other agencies consider in determining whether an IC controls how the work is done. (See Chapters 2 through 6 for information on these factors.)

When you draft your own agreement, include only those provisions that apply to your situation. The more that apply, the more likely the worker will be viewed as an IC.

To obtain safe harbor protection from the IRS, you must have a reasonable basis for classifying the workers involved as ICs. One reasonable basis is that such classification is a longstanding practice in your trade or industry and that you relied on this longstanding practice in making your classification. (See Chapter 3 for more about safe harbor protection.)

If you are relying on industry practice, you should say so in your agreement.

Suggested Language

Independent Contractor Status

Contractor is an independent contractor, not Client's employee. Contractor's employees or contract personnel are not Client's employees. Contractor and Client agree to the following rights and obligations consistent with an independent contractor relationship:

- Contractor has the right to perform services for others during the term of this Agreement.

- Contractor has the sole right to control and direct the means, manner, and method by which the services required by this Agreement will be performed.

- Contractor has the right to perform the services required by this Agreement at any place, location, or time.

- Contractor will furnish all equipment and materials used to provide the services required by this Agreement.

- Contractor has the right to hire assistants as subcontractors, or to use employees to provide the services required by this Agreement.

- Contractor or Contractor's employees or contract personnel shall perform the services required by this Agreement; Client shall not hire, supervise, or pay any assistants to help Contractor.

- Neither Contractor nor Contractor's employees or contract personnel shall receive any training from Client in the skills necessary to perform the services required by this Agreement.

- Client shall not require Contractor or Contractor's employees or contract personnel to devote full time to performing the services required by this Agreement.

Optional Addition

[OPTIONAL:] Contractor acknowledges that Contractor has been classified as an independent contractor because such classification is a long-standing practice of a significant segment of Client's trade or industry.

Business Permits, Certificates, and Licenses

The IC should have all business permits, certificates, and licenses needed to perform the work. For example, if you hire an IC to perform construction work, the IC should have a contractor's license if one is required by your state's law. A worker who doesn't have required licenses and permits looks like an employee, not an IC in business for himself or herself. The IC should obtain such licenses and permits on his or her own; you should not pay for them.

Suggested Language

Business Permits, Certificates, and Licenses
Contractor has complied with all federal, state, and local laws requiring business permits, certificates, and licenses to carry out the services to be performed under this Agreement.

State and Federal Taxes

Do not pay or withhold any taxes on an IC's behalf. Doing so is a very strong indicator that the worker is an employee. Indeed, some courts have held that workers were employees based on this factor alone.

Include a straightforward provision, such as the one suggested below, to make sure the IC understands that he or she must pay all applicable taxes.

Suggested Language

State and Federal Taxes

Client will not:

- withhold FICA (Social Security and Medicare taxes) from Contractor's payments or make FICA payments on Contractor's behalf
- make state or federal unemployment compensation contributions on Contractor's behalf, or
- withhold state or federal income tax from Contractor's payments.

Contractor shall pay all taxes incurred while performing services under this Agreement—including all applicable income taxes and, if Contractor is not a corporation, self-employment (Social Security) taxes. Upon demand, Contractor shall provide Client with proof that such payments have been made.

Fringe Benefits

Do not provide ICs or their employees with fringe benefits that you provide to your own employees, such as health insurance, pension benefits, child care allowances, or even the right to use employee facilities, such as an exercise room.

If the IRS or another government agency audits you and concludes the worker should have been classified as your employee, the worker may become entitled to benefits you provide to your employees such as pension or health benefits. The following clause includes language by which the contractor waives (gives up) any right to such benefits.

Suggested Language

Fringe Benefits

Contractor understands that neither Contractor nor Contractor's employees or contract personnel are eligible to participate in any employee pension, profit sharing, health, vacation pay, sick pay, or other fringe benefit plan of Client. If Contractor is later classified as Client's employee, Contractor expressly waives Contractor's rights to any benefits to which Contractor was, or might have become entitled.

Workers' Compensation

If a worker qualifies as an IC under your state workers' compensation law, do not provide the IC with workers' compensation coverage.

If the IC has employees, the IC should provide them with workers' compensation coverage. The only exception would be where the IC is located in one of 13 states that don't require workers' comp coverage for employers that have fewer than a designated number of employees—anywhere from three to five depending on the state. (See Chapter 6 for details.)

If an IC's employees lack workers' compensation coverage, your own workers' compensation insurer will probably require you to provide coverage under your own policy and pay an additional workers' compensation premium. To avoid this, require ICs to provide you with a certificate of insurance establishing that an IC's employees are covered by workers' compensation insurance. A certificate of insurance issued by the workers' compensation insurer is written proof that the IC has a workers' compensation policy.

Whether an IC has employees or not, you may want to require the worker to have his or her own workers' compensation coverage. Self-employed people who have no employees are usually not required to have workers' compensation coverage, but they usually can obtain a low-cost "if any" worker's comp policy. More and more hiring firms are imposing this requirement because their own workers' compensation insurers may require them to cover ICs who don't have their own workers' compensation insurance. (See Chapter 6 for more about workers' compensation issues and ICs.)

Suggested Language

> **Workers' Compensation**
> Client shall not obtain workers' compensation insurance on behalf of Contractor or Contractor's employees. Contractor shall provide its employees and agents with workers' compensation insurance coverage before they perform any services under this Agreement. Contractor will provide Client with a certificate of workers' compensation insurance before such services are commenced. Contractor agrees to hold harmless and indemnify Client for any and all claims arising out of any injury, disability, or death of any of Contractor's employees or agents.

Optional Addition

> **Workers' Compensation**
> Contractor shall obtain an "if any" policy of workers' compensation insurance coverage. Contractor shall provide Client with proof that such coverage has been obtained before starting work.

Use this optional language if the IC claims to have no employees and you want him or her to obtain an "if any" workers' compensation policy that provides coverage if the IC is later determined to have employees.

Unemployment Compensation

If the worker qualifies as an IC under your state's unemployment compensation law, do not pay unemployment compensation taxes for him or her. (See Chapter 5 for more about unemployment issues and ICs.) The IC will not be entitled to receive unemployment compensation benefits when the work is finished or the agreement is terminated.

Suggested Language

Unemployment Compensation

Client shall make no state or federal unemployment compensation payments on behalf of Contractor or Contractor's employees or contract personnel. Contractor will not be entitled to these benefits in connection with work performed under this Agreement.

Insurance

An IC should have his or her own liability insurance policy just like any other business; you need not provide it. This type of coverage insures the IC against personal injury or property damage claims by others. For example, if an IC accidentally injures a bystander while performing services for you, the IC's liability policy will pay the costs of defending a lawsuit and pay damages up to the limits of the policy coverage. This helps eliminate an injured person's motivation to attempt to recover from you as well as the IC for fear the IC won't be able to pay. Also, having insurance helps show that the IC is in business. One million dollars is a standard minimum amount of liability coverage often required.

The IC should also agree to indemnify—that is, repay—you if somebody he or she injures decides to sue you.

Suggested Language

Insurance

Client shall not provide any insurance coverage of any kind for Contractor or Contractor's employees or contract personnel. Contractor agrees to maintain an insurance policy of at least $[*state amount*] to cover any negligent acts committed by Contractor or Contractor's employees or agents while performing services under this Agreement.

Contractor shall indemnify and hold Client harmless from any loss or liability arising from performing services under this Agreement.

Terminating the Agreement

An independent contractor agreement should contain a provision establishing the circumstances under which it may be terminated. Because an IC is not an employee, you should not have the right to "fire" or terminate an IC for any reason or no reason at all as you would with an employee under the generally prevailing employment-at-will doctrine. Likewise, the IC should not be able simply to "quit" at any time the same way an at-will employee can with no legal liability.

Instead, your agreement should require you and the IC to provide at least some notice before any termination becomes effective. However, such notice is not required if the IC—or hiring firm—is in "material breach" of the agreement. A material breach is a serious violation of the agreement— for example, the IC's failure to produce results or meet the deadlines specified in the agreement, or the hiring firm's failure to pay the IC.

It's up to you and the IC to decide how many days' notice will have to be provided to terminate the agreement without "cause." Thirty days is a common term, but a shorter time period can be used—for example, seven or 14 days.

The hiring firm will be liable for paying for work the IC completed before the agreement was terminated, but not anything the IC did after the effective date of termination.

Suggested Language

Terminating the Agreement

Either party may terminate this Agreement:

- without cause, by [*number*] days' prior written notice, or
- with cause, immediately upon material breach of any term of this Agreement by the other party.

Client shall promptly pay Contractor for services performed before the effective date of termination. Client shall not owe Contractor compensation for any services performed following such date.

Exclusive Agreement

Business contracts typically contain a provision stating that the written agreement is the complete and exclusive agreement between those involved. This reinforces the idea that what is written in the contract is the entire agreement. Neither you nor the IC can later bring up side letters, oral statements, or other material not covered by the contract. A clause such as this avoids later claims that promises not contained in the written contract were made and broken.

Because of this provision, you must make sure that any promises or statements the IC made are included in the agreement if you consider them to be part of the deal. This may include the IC's proposal or bid, sales literature, side letters, and so forth. If they aren't included in or attached to the contract, they won't be considered part of the agreement.

Suggested Language

Exclusive Agreement
This is the entire Agreement between Contractor and Client.

Severability

The following standard contract provision permits the agreement as a whole to continue even if portions of it are found invalid by a court or an arbitrator. For example, if, for some reason, a court found the provision requiring the IC to obtain workers' compensation coverage to be invalid, the rest of the contract would remain in force.

Suggested Language

Severability
If any provision of this Agreement is determined to be invalid, illegal, or unenforceable, the remaining provisions will remain in full force if the essential provisions of the Agreement for Client and Contractor remain valid, binding, and enforceable.

Applicable Law and Forum

Each state has its own set of laws, and you will have to look to these laws if you and the IC ever have a dispute over the agreement. If you and the IC have offices in the same state, that state's law will apply. But if your offices are in different states, you'll need to decide which state's law should govern the agreement. There is some advantage to having the law of your own state govern, because your local attorney will be more familiar with that law.

The agreement should also contain a jurisdiction provision (also called a forum selection clause). This requires the parties to agree in advance as to where a legal case can be filed if there is a dispute. This can be important, especially if the parties are located far apart. If your office is in Oregon and your IC is in Florida, do you want to have to go to Florida to file a lawsuit if the IC fails to perform? Probably not.

Often, both the applicable law and jurisdiction provisions are combined as in the following suggested clause.

Suggested Language

> **Applicable Law and Forum**
>
> This Agreement will be governed by the laws of the state of [*indicate state in which you have your main office*], and any disputes arising from it must be litigated exclusively in the federal or state courts located in [*insert county and state in which parties agree to litigate*].

Notices

When you want to do something important involving the agreement— terminate it, for example—you need to let the IC know. This is called giving notice. The following provision gives you several options for providing the IC with notice: by personal delivery, by mail, or by fax or email followed by a confirming letter.

If you give notice by mail, it is not effective until three days after it's sent. For example, if you want to end the agreement on 30-days' notice and mail your notice of termination to the IC, the agreement will not end until 33 days after you mailed it.

The following provision requires that if notice is given by email, it must be followed up by notice by postal mail. Relying solely on notice by email can be dangerous—email can be easily overlooked or accidentally deleted by the recipient.

If the notice procedure set forth in the clause below is followed, notice is effective. This means that, for legal purposes, the recipient has received notice, whether he or she actually got it or not. This prevents the recipient from avoiding the notice in order to claim he or she never got it.

Suggested Language

Notices

All notices and other communications in connection with this Agreement shall be in writing and shall be considered effective only as follows:

- when delivered personally to the recipient's address as stated on this Agreement
- three days after being deposited in the United States mail, with postage prepaid to the recipient's address as stated on this Agreement, or
- when sent by fax or email to the last fax number or email address of the recipient known to the person giving notice. Notice is effective upon receipt provided that a duplicate copy of the notice is promptly given by first class mail or the recipient delivers a written confirmation of receipt.

No Partnership

You want to make sure that you and the IC are considered separate legal entities, not partners. If an IC is viewed as your partner, you'll be liable for his or her debts, and the IC will have the power to make contracts that obligate you to others without your consent.

Suggested Language

No Partnership
This Agreement does not create a partnership relationship. Contractor does not have authority to enter into contracts on Client's behalf.

Assignment

Assignment is a catchall term used to describe two different things: an assignment of rights and a delegation of duties. An assignment is the process by which rights or benefits under a contract are transferred to someone else. For example, the IC might assign to someone else the right to be paid the money due for performing services for you. Delegation is the flipside of assignment. Instead of transferring benefits under a contract, a delegation transfers the duties. For example, an IC may delegate to someone else the duty to perform the services for which you've contracted.

Delegation of the IC's duties is the most important issue here. You've hired the IC to do the job, but the IC might try to get someone else to do it in his or her place. Fortunately, the law protects you against an IC who tries to fob off the work on someone else. If the work the IC has agreed to do involves personal services, the IC cannot delegate it to someone else without your consent.

Examples of personal services include those by professionals such as lawyers, physicians, or architects. When you hire an architect, he or she can't pass your project to another architectural firm without your consent. Courts consider it unfair for either a client or an IC to change horses in midstream. Services by IC writers, artists, and musicians are also considered too personal to be delegated without the hiring firm's consent.

Service contracts with ICs involving more mechanical tasks ordinarily are delegable without the hiring firm's consent. For example, a contract to construct or paint a house would ordinarily be delegable. However, even in these cases, the IC can't delegate his or her duties to a person who lacks the skill or experience to complete the work satisfactorily.

Your agreement need not mention assignment and delegation. The work will be assignable or delegable subject to the restrictions discussed above. This is the best option in most cases.

Assignment Allowed

If you absolutely do not care who does the work required by the agreement, choose Alternative A to allow an unrestricted right of assignment and delegation.

This clause can help you if you are challenged for classifying the worker as an IC. It is strong evidence of an IC relationship because it shows you're concerned only with results, not who achieves them. It also demonstrates your lack of control over the IC.

Note, however, that if you include this clause in the agreement, the restrictions on assignment and delegation discussed above will not apply— that is, the IC will be able to delegate his or her duties even if they involve personal services.

Suggested Language: Alternative A

Assignment
Either Contractor or Client may assign, delegate, or subcontract any rights or obligations under this Agreement.

Approval Required

There may be some situations in which you really don't want an IC to assign his or her contractual duties without your consent. This will often be the case if you hire a particular IC because of his or her special expertise, reputation for performance, or financial stability.

Suggested Language: Alternative B

Assignment
Contractor may not assign, delegate, or subcontract any rights or obligations under this Agreement without Client's prior written approval.

Choose Alternative A to allow the IC to assign contractual rights and obligations to others, or Alternative B if you wish to restrict the IC's right to assign contractual rights and obligations.

Resolving Disputes

If you and the IC get into a dispute—for example, over the price or quality of the IC's services—it is best if the two of you can resolve the matter through informal negotiations. This is by far the easiest and cheapest way to resolve any problem.

Unfortunately, informal negotiations don't always work. If you and the IC reach an impasse, you may need help. The following part of the agreement describes how you will handle any disputes you can't resolve on your own. Three alternatives are provided: mediation, arbitration, and court.

Mediation

Choose Alternative A if you want to try mediation before going to court. Mediation, an increasingly popular alternative to going to court, works like this: You and the IC choose a neutral third person to try to help you settle your dispute. The mediator has no power to impose a decision, but will try to help you arrive at one. Unless both parties agree to the resolution, there is no resolution.

Insert the place where the mediation meeting will occur. You'll usually want it in the city or county where your office is located. You don't want to have to travel a long distance to attend a mediation.

If the mediation doesn't help resolve the dispute, you still have the option of going to court. If you wish, you may include the optional clause requiring the loser in any lawsuit to pay the other person's attorneys' fees.

Suggested Language: Alternative A

Resolving Disputes

If a dispute arises under this Agreement, the parties agree to first try to resolve the dispute with the help of a mutually agreed-upon mediator in [*insert city or county, and state in which parties agree to mediate*]. Any costs and fees other than attorneys' fees associated with the mediation shall be shared equally by the parties.

If the dispute is not resolved within 30 days after it is referred to the mediator, any party may take the matter to court.

[*OPTIONAL:*] If any court action is necessary to enforce this Agreement, the prevailing party shall be entitled to reasonable attorneys' fees, costs, and expenses, in addition to any other relief to which he or she may be entitled.

Arbitration

Choose Alternative B if you want to avoid court altogether. Under this clause, you and the IC first try to resolve your dispute through mediation. If this doesn't work, you must submit the dispute to binding arbitration. Arbitration is like an informal court trial without a jury; instead of a judge, an arbitrator (or panel of arbitrators) chosen by the parties decides the outcome.

You and the IC can agree on anyone to serve as the arbitrator. Arbitrators are often retired judges, lawyers, or people with special expertise in the field involved. Businesses often use private dispute resolution services that maintain a roster of arbitrators. The best known of these is the American Arbitration Association, which has offices in most major cities.

You may be represented by a lawyer in the arbitration, but it's not required. The arbitrator's decision is final and binding—that is, you can't go to court and try the dispute again if you don't like the arbitrator's decision, except in unusual cases where the arbitrator has been guilty of fraud, misconduct, or bias.

By using this provision, you give up your right to go to court. The advantage is that arbitration is usually much cheaper and faster than court litigation.

This provision states that the arbitrator's award can be converted into a court judgment. This means that if the losing side doesn't pay the money required by the award, the other party can easily obtain a judgment and enforce it like any other court judgment—for example, have the losing side's bank accounts and property seized to pay the amount due.

The provision leaves it up to the arbitrator to decide who should pay the costs and fees associated with the arbitration.

You must insert the place where the arbitration hearing will occur. For your own convenience, you'll usually want it in the city or county where your office is located.

Suggested Language: Alternative B

Resolving Disputes

If a dispute arises under this Agreement, the parties agree to first try to resolve the dispute with the help of a mutually agreed-upon mediator in [*insert city or county, and state in which parties agree to mediate*]. Any costs and fees other than attorneys' fees associated with the mediation shall be shared equally by the parties.

If it proves impossible to arrive at a mutually satisfactory solution through mediation, the parties agree to submit the dispute to a mutually agreed-upon arbitrator in [*insert city or county, and state in which parties agree to mediate*]. Judgment upon the award rendered by the arbitrator may be entered in any court having jurisdiction to do so. Costs of arbitration, including attorneys' fees, will be allocated by the arbitrator.

RESOURCE

Want to know more about alternatives to court? For a detailed discussion of mediation and arbitration, see *Mediate, Don't Litigate: Strategies for Successful Mediation,* by Peter Lovenheim and Lisa Guerin (Nolo), an eBook available at www.nolo.com.

Court Litigation

Choose Alternative C if you want to resolve disputes by going to court. This is the traditional way contract disputes are resolved. It is also usually the most expensive and time-consuming.

The optional language provides that if either person has to sue the other in court to enforce the agreement and wins, the loser is required to pay the other person's attorneys' fees and expenses. Without this clause, each side must pay its own expenses. This clause can help make filing a lawsuit economically feasible and will give the IC an additional reason to settle if you have a strong case.

However, there may be situations in which you do not want to include an attorneys' fees provision. An IC who has little or no money won't be able to pay the fees, so the provision won't do you much good. What's worse, an attorneys' fees provision could help the IC convince a lawyer to file a case against you, because the lawyer will be looking to you for his or her fee instead of asking the IC to provide a cash retainer up front. If you have substantially more financial resources than the IC or think it's more likely you'll break the contract than the IC will, an attorneys' fees provision is not in your interests—and you should not include the optional clause in your agreement.

Suggested Language: Alternative C

Resolving Disputes

If a dispute arises under this Agreement, any party may take the matter to court.

[OPTIONAL:] If any court action is necessary to enforce this Agreement, the prevailing party shall be entitled to reasonable attorneys' fees, costs, and expenses, in addition to any other relief to which he or she may be entitled.

Choose Alternative A to require mediation, Alternative B to require arbitration, or Alternative C if you wish to require court action to enforce the agreement.

Signatures

The end of the main body of the agreement should contain spaces for you to sign, write your title, and date, and in which the IC can also sign and provide a taxpayer ID number.

> **CAUTION**
>
> **Don't forget the numbers.** Be sure to obtain the IC's Social Security number, or taxpayer ID number if the IC is a corporation or partnership. If you don't, you'll have to withhold federal income taxes from the IC's pay. (See Chapter 11 for more about this issue.)

Suggested Language

Signatures

Client: _____
Name of Client

By: _____
Signature

_____
Typed or Printed Name

Title: _____

Date: _____

Contractor: _____
Name of Contractor

By: _____
Signature

_____
Typed or Printed Name

Title: _____

Taxpayer ID Number: _____

Date: _____

Electronic and Digital Signatures

It is common, and perfectly legal, to use electronic or digital signatures in a contract instead of handwritten signatures. But it is a good idea to include a provision at the end of the agreement making it clear that such signatures may be used.

Optional Language

Electronic or Digital Signatures
This agreement may be signed by an electronic or digital signature.

Optional Provisions

The following provisions are not absolutely necessary for every IC agreement, but you may want to include one or more in your agreement depending on the circumstances.

Modifying the Agreement

No contract is engraved in stone. You and the IC can always modify or amend your contract if circumstances change. You can even agree to call the whole thing off.

EXAMPLE: Barbara, an IC well digger, agrees to dig a 50-foot-deep well on property owned by Kate for $2,000. After digging ten feet, Barbara hits solid rock that no one knew was there. To complete the well, she'll have to lease expensive, heavy equipment. To defray the added expense, she asks Kate to pay her $4,000 instead of $2,000 for the work. Kate agrees. Barbara and Kate have amended their original agreement.

Neither you nor the IC is ever obligated to accept a proposed modification to your contract. Either of you can always say no and accept the consequences, which at their most dire may mean a court battle over breaking the original contract. However, you're usually better off reaching some sort of accommodation with the IC, unless he or she is totally unreasonable.

Unless your contract is one that must be in writing to be legally valid—most commonly, an agreement that can't be performed in less than one year—or is in writing and explicitly states it may only be modified in writing, it can usually be modified by an oral agreement. In other words, you need not write down the changes.

> **EXAMPLE:** Art signs a contract with Zeno to build an addition to his house. Halfway through the project, Art decides that he wants Zeno to do some extra work not covered by their original agreement. Art and Zeno have a telephone conversation in which Zeno agrees to do the extra work for extra money. Although nothing is put in writing, their change to their original agreement is legally enforceable.

In the real world, people make changes to their contracts all the time without writing them down. The flexibility afforded by such an informal approach to contract amendments might be just what you want. However, you should be aware that misunderstandings and disputes can arise from this approach. It's always best to have something in writing showing what you've agreed to do. You can do this informally. For example, you can simply send a confirming email or letter following a telephone call with an IC summarizing the changes you both agreed to make. Be sure to keep a copy for your files. Or if the amendment involves a contract provision that is very important—the IC's payment, for example—you can insist on a written amendment signed by you and the IC.

You may also add a provision to the contract requiring that all amendments be in writing and signed by you and the IC before they become effective. Courts are often reluctant to enforce provisions that prohibit oral amendments, but they can still be useful to remind the parties that they should put any contract changes in writing. You can still negotiate changes by phone or email, but you should write them down.

Modifying the Agreement

This Agreement may be amended only by a writing signed by both Client and Contractor.

Work at Your Premises

An IC should not work at your office or other premises you maintain unless the nature of the work absolutely requires it. For example, a computer consultant may have to perform work on your computers at your office. In these situations, it's a good idea to add a provision to the agreement making it clear that the IC is working at your premises because the work requires it.

Work at Your Premises

Because of the nature of the services to be provided by Contractor, Client agrees to furnish space on its premises for Contractor while performing these services.

Indemnification

If the IRS or another government agency determines that you have misclassified a worker as an IC, you may have to pay back taxes, fines, and penalties. This is one of the greatest risks of hiring ICs. Some hiring firms try to shift this risk to the IC's shoulders by including an indemnification clause in their IC agreements. Such a provision requires the IC to repay the hiring firm for any losses it suffers if the IC is reclassified as an employee.

This may sound attractive at first, but there are several reasons why it's usually not a good idea to include such a clause in an IC agreement:

- You are prohibited by law from recovering from a worker any taxes and penalties the IRS assesses against you if it determines that you intentionally misclassified the worker as an IC.
- The clause is practically useless if you're unable to locate the IC when you're audited or if the IC doesn't have the money to repay you for your losses.
- The clause makes it look as if you're not sure whether the worker is really an IC, and it may cause an auditor to have similar doubts.
- Many intelligent ICs will refuse to sign an agreement containing such a clause.
- The Department of Labor does not permit employers to use such clauses to shift liability for failure to pay workers overtime.
- Even if you can locate the IC and he or she has the money to repay you for your losses, you'll still have to spend time and money to go to court if the IC refuses to pay. There is a chance you'll lose because a court might conclude that the indemnification clause goes against public policy and is unenforceable for some of the reasons stated above. This is why few hiring firms ever try collecting on indemnification provisions.

Intellectual Property Ownership

If you hire an IC to create intellectual property—for example, writings, music, software programs, designs, or inventions—you should include a provision securing the rights to the finished product. Unless you include a specific provision about the assignment or transfer of intellectual property rights to you, you can never be sure that you will own the work you pay the IC to create.

You can use an assignment or a work-made-for-hire agreement.

Assignment

By including the following clause in your agreement, the IC assigns to you his or her rights in any intellectual property created on your behalf. This gives you full ownership, but the IC retains the right to terminate the assignment 35 years after it is made and get back his or her copyrights for nothing. (See Chapter 9 for a detailed discussion.)

Suggested Language A

Intellectual Property Ownership

Contractor assigns to Client all patent, copyright, trade secret, and other intellectual property rights in anything created or developed by Contractor for Client under this Agreement. Contractor shall help prepare any documents Client considers necessary to secure any copyright, patent, trade secret, and other intellectual property rights at no charge to Client. However, Client shall reimburse Contractor for reasonable out-of-pocket expenses.

Having paid the IC to create intellectual property for you, you probably won't want the IC to use the material for others without your permission and will probably want to include the following paragraph. However, this is always subject to negotiation.

Optional Addition

Intellectual Property Ownership

Contractor agrees not to use any of the intellectual property mentioned above for the benefit of any other party without Client's prior written permission.

Work-Made-for-Hire Agreement

As an alternative to an assignment, you can enter into a work-made-for-hire agreement with the IC. By using such an agreement, you are considered the author of any copyrightable work the IC creates. This gives you all the copyrights rights in the works created, and the IC has no ability to terminate the transfer 35 years later as is the case with an assignment. However, clients who use a work-made-for-hire agreement in California should familiarize themselves with state law (see Chapter 9 for a detailed discussion) because the agreement may result in the worker's being categorized as an employee.

Also, keep in mind that not all works of authorship can be classified as works-made-for-hire. For that reason, the clause below provides an assignment as an alternative to the work-made-for-hire arrangement.

Suggested Language B

Intellectual Property Ownership

To the extent that the work performed by Contractor under this Agreement (Contractor's Work) includes any work of authorship entitled to protection under copyright law, the parties agree to the following provisions:

- Contractor's Work has been specially ordered and commissioned by Client as a contribution to a collective work, a supplementary work, or other category of work eligible to be treated as a work made for hire under the United States Copyright Act.

- Contractor's Work shall be deemed a commissioned work and a work made for hire to the greatest extent permitted by law.

- Client shall be the sole author of Contractor's Work and any work embodying Contractor's Work according to the United States Copyright Act.

- To the extent that Contractor's Work is not properly characterized as a work made for hire, Contractor assigns and grants to Client all rights, title, and interest in Contractor's Work, including all copyright rights, in perpetuity and throughout the world.

- Contractor shall help prepare any papers Client considers necessary to secure any copyrights, patents, trademarks, or other intellectual property rights at no charge to Client. However, Client shall reimburse Contractor for reasonable out-of-pocket expenses.

- Contractor agrees to require any employees or contract personnel Contractor uses to perform services under this Agreement to assign in writing to Contractor all copyright and other intellectual property rights they may have in their work product. Contractor shall provide Client with a signed copy of each such assignment.

Confidentiality

If, during the course of his or her work, an IC may have access to your valuable trade secrets—for example, customer lists, business plans, business methods, and techniques not known by your competitors—it is reasonable for you to include a nondisclosure provision in the agreement. Such a provision prohibits the IC from disclosing your trade secrets to others without your permission. (See Chapter 9 for more about trade secrets.)

Suggested Language

Confidentiality

Contractor will not disclose or use, either during or after the term of this Agreement, any proprietary or confidential information of Client without Client's prior written permission except to the extent necessary to perform services on Client's behalf.

Proprietary or confidential information includes:

- the written, printed, graphic, or electronically recorded materials furnished by Client for Contractor to use
- business plans, customer lists, operating procedures, trade secrets, design formulas, know-how and processes, computer programs and inventories, discoveries, and improvements of any kind, and
- information belonging to customers and suppliers of Client about which Contractor gained knowledge as a result of Contractor's services to Client.

Contractor shall not be restricted in using any material that is publicly available, that is already in Contractor's possession or known to Contractor without restriction, or that is rightfully obtained by Contractor from sources other than Client.

Suggested Language (Confidentiality Continued)

Upon termination of Contractor's services to Client, or at Client's request, Contractor shall deliver to Client all materials in Contractor's possession relating to Client's business.

These confidentiality obligations survive termination of this Agreement.

Nonsolicitation

If you're concerned about an IC getting to know your clients or customers and perhaps stealing business away from you, you can include the following nonsolicitation clause in the agreement.

Suggested Language

Nonsolicitation

For a period of [*fill in period from two months to three years*] after termination of this Agreement, Contractor agrees not to call on, solicit, or take away Client's customers or potential customers of which Contractor became aware as a result of Contractor's services for Client.

Sample IC Agreement

The provisions in the sample agreement below are discussed above. You should be able to craft some form of this agreement to meet your needs if you hire any type of general independent contractor.

 FORM
You can download a copy of this agreement on the Nolo website; the link is included in Appendix A.

Independent Contractor Agreement

This Agreement is made between Acme Widget Co. (Client), with a principal place of business at 123 Main Street, Marred Vista, CA 90000, and ABC Consulting, Inc. (Contractor), with a principal place of business at 456 Grub Street, Santa Longo, CA 90001.

1. Term of Agreement

This Agreement will become effective on May 1, 20xx, and will end no later than the earlier of June 1, 20xx and completion of Contractor's services under this Agreement.

2. Services to Be Performed

Contractor agrees to perform the following services: Install and test Client's DX9-105 widget manufacturing press so that it performs according to the manufacturer's specifications.

3. Payment

In consideration for the services to be performed by Contractor, Client agrees to pay Contractor $20,000 according to the terms set out below.

4. Terms of Payment

Upon completing Contractor's services under this Agreement, Contractor shall submit an invoice. Client shall pay Contractor the compensation described within 30 days after receiving Contractor's invoice.

5. Expenses

Contractor shall be responsible for all expenses incurred while performing services under this Agreement. These include license fees, memberships, and dues; automobile and other travel expenses; meals and entertainment; insurance premiums; and all salary, expenses, and other compensation paid to employees or contract personnel Contractor hires to complete the work under this Agreement.

6. Independent Contractor Status

Contractor is an independent contractor, not Client's employee. Contractor's employees or contract personnel are not Client's employees. Contractor and Client agree to the following rights and obligations consistent with an independent contractor relationship:

- Contractor has the right to perform services for others during the term of this Agreement.
- Contractor has the sole right to control and direct the means, manner, and method by which the services required by this Agreement will be performed.
- Contractor has the right to perform the services required by this Agreement at any place, location, or time.
- Contractor will furnish all equipment and materials used to provide the services required by this Agreement.
- Contractor has the right to hire assistants as subcontractors, or to use employees to provide the services required by this Agreement.
- Contractor or Contractor's employees or contract personnel shall perform the services required by this Agreement; Client shall not hire, supervise, or pay any assistants to help Contractor.
- Neither Contractor nor Contractor's employees or contract personnel shall receive any training from Client in the skills necessary to perform the services required by this Agreement.
- Client shall not require Contractor or Contractor's employees or contract personnel to devote full time to performing the services required by this Agreement.

7. Business Permits, Certificates, and Licenses

Contractor has complied with all federal, state, and local laws requiring business permits, certificates, and licenses to carry out the services to be performed under this Agreement.

8. State and Federal Taxes

Client will not:

- withhold FICA (Social Security and Medicare taxes) from Contractor's payments or make FICA payments on Contractor's behalf
- make state or federal unemployment compensation contributions on Contractor's behalf, or
- withhold state or federal income tax from Contractor's payments.

Contractor shall pay all taxes incurred while performing services under this Agreement—including all applicable income taxes and, if Contractor is not a corporation, self-employment (Social Security) taxes. Upon demand, Contractor shall provide Client with proof that such payments have been made.

9. Fringe Benefits

Contractor understands that neither Contractor nor Contractor's employees or contract personnel are eligible to participate in any employee pension, profit sharing, health, vacation pay, sick pay, or other fringe benefit plan of Client. If Contractor is later classified as Client's employee, Contractor expressly waives Contractor's rights to any benefits to which Contractor was, or might have become, entitled.

10. Workers' Compensation

Client shall not obtain workers' compensation insurance on behalf of Contractor or Contractor's employees. Contractor shall provide its employees and agents with workers' compensation insurance coverage before they perform any services under this Agreement. Contractor will provide Client with a certificate of workers' compensation insurance before such services are commenced. Contractor agrees to hold harmless and indemnify Client for any and all claims arising out of any injury, disability, or death of any of Contractor's employees or agents.

11. Unemployment Compensation

Client shall make no state or federal unemployment compensation payments on behalf of Contractor or Contractor's employees or contract personnel. Contractor will not be entitled to these benefits in connection with work performed under this Agreement.

12. Insurance

Client shall not provide any insurance coverage of any kind for Contractor or Contractor's employees or contract personnel. Contractor agrees to maintain an insurance policy of at least $1 million to cover any negligent acts committed by Contractor or Contractor's employees or agents while performing services under this Agreement.

Contractor shall indemnify and hold Client harmless from any loss or liability arising from performing services under this Agreement.

13. Terminating the Agreement

Either party may terminate this Agreement:

- without cause, by seven days' prior written notice, or
- with cause, immediately upon material breach of any term of this Agreement by the other party.

Client shall promptly pay Contractor for services performed before the effective date of termination. Client shall not owe Contractor compensation for any services performed following such date.

14. Exclusive Agreement

This is the entire Agreement between Contractor and Client.

15. Severability

If any provision of this Agreement is determined to be invalid, illegal, or unenforceable, the remaining provisions will remain in full force if the essential provisions of this Agreement for Client and Contractor remain valid, binding, and enforceable.

16. Applicable Law and Forum

This Agreement will be governed by the laws of the state of California, and any disputes arising from it must be litigated exclusively in the federal or state courts located in Alameda County, California.

17. Notices

All notices and other communications in connection with this Agreement shall be in writing and shall be considered effective only as follows:

- when delivered personally to the recipient's address as stated on this Agreement
- three days after being deposited in the United States mail, with postage prepaid to the recipient's address as stated on this Agreement, or
- when sent by fax or email to the last fax number or email address of the recipient known to the person giving notice. Notice is effective upon receipt provided that a duplicate copy of the notice is promptly given by first class mail or the recipient delivers a written confirmation of receipt.

18. No Partnership

This Agreement does not create a partnership relationship. Contractor does not have authority to enter into contracts on Client's behalf.

19. Assignment

Either Contractor or Client may assign, delegate, or subcontract any rights or obligations under this Agreement.

20. Resolving Disputes

If a dispute arises under this Agreement, the parties agree to first try to resolve the dispute with the help of a mutually agreed-upon mediator in Alameda County, California. Any costs and fees other than attorneys' fees associated with the mediation shall be shared equally by the parties.

If the dispute is not resolved within 30 days after it is referred to the mediator, any party may take the matter to court.

If any court action is necessary to enforce this Agreement, the prevailing party shall be entitled to reasonable attorneys' fees, costs, and expenses, in addition to any other relief to which he or she may be entitled.

21. Signatures

Client: Acme Widget Co.

By: _Basilio Chew_____
Signature

_Basilio Chew_____
Typed or Printed Name

Title: _President_____

Date: _April 30, 20xx_____

Contractor: ABC Consulting, Inc.

By: _George Bailey_____
Signature

_George Bailey_____
Typed or Printed Name

Title: _President_____

Taxpayer ID Number: _123-45-6789_____

Date: _April 30, 20xx_____

Electronic or Digital Signatures

This agreement may be signed by an electronic or digital signature.

Help Beyond This Book

Finding and Using a Lawyer 334

    What Type of Lawyer Do You Need? 334

    Finding a Lawyer 335

Help From Other Experts 336

    Tax Professionals 336

    Insurance Brokers 336

    Industry and Trade Associations 337

Doing Your Own Legal Research 337

    Researching Federal Tax Law 337

    Researching Other Areas of Law 341

    Online Resources 342

The legal issues involved in hiring ICs are complex and varied. You may have questions that aren't answered by this book. This chapter provides guidance on how to find and use more specific legal resources, including lawyers and other knowledgeable experts. It also explains the basics of doing your own legal research.

Finding and Using a Lawyer

An experienced attorney may help answer your questions and allay your fears about working with independent contractors.

What Type of Lawyer Do You Need?

Many different areas of law may apply when you hire ICs, including:
- federal tax law
- state tax law
- workers' compensation law
- federal and state employment and antidiscrimination laws, and
- general business law.

Unfortunately, you may find it difficult or impossible to find a single attorney to competently advise you about all these legal issues. For example, an attorney who knows the fine points of tax law may know nothing about your state's workers' compensation and employment laws.

However, there are attorneys who specialize in advising businesses, both large and small. And some of them have experience dealing with the legal issues faced by businesses that hire ICs. These lawyers are a bit like general practitioner doctors: They know a little about a lot of different areas of law.

A lawyer with plenty of experience working with businesses like yours may know enough to answer your questions. For example, if you run a software company, a lawyer who customarily represents software firms may be familiar with the laws as they apply to the software industry and should also know about the legal problems that other software companies that hire ICs have encountered.

But if you can't find a lawyer with this type of experience, you may need to consult more than one attorney or consult other types of experts. For example, a CPA may be able to help you with questions about IRS procedures, while a business lawyer can help you deal with unemployment compensation, employment, and other state laws.

Hiring a Tax Attorney to Handle IRS Audits

If you're facing an IRS field audit that could result in significant assessments or penalties, it makes sense to hire a tax attorney. A tax attorney is a lawyer with either a special tax law degree, an LL.M. in taxation, or a tax specialization certification from a state bar association. Tax attorneys specialize in representing taxpayers before the IRS and in court.

Finding a Lawyer

When you begin looking for a lawyer, try to find someone with experience representing businesses similar to yours.

The best way to locate a lawyer is through referrals from other businesses in your community that use ICs. Industry associations and trade groups are also excellent sources of referrals. If you already have or know a lawyer, he or she might also be able to refer you to an experienced person who has the qualifications you need. Other people, such as your banker, accountant, or insurance agent may also know of good business lawyers.

RESOURCE

Where else to look for a lawyer? Another tried-and-true method of finding an attorney is through a directory. Nolo's Lawyer Directory, at www.nolo.com/lawyers, provides not only attorneys organized by practice area, but also extensive lawyer profiles.

Help From Other Experts

Lawyers aren't the only ones who can help you deal with the legal issues involved in hiring ICs. Tax professionals, insurance brokers, and trade groups can also be very helpful.

Tax Professionals

Attorneys are usually the most expensive, but not always the most knowledgeable, professionals you can go to for advice on tax law. You can get outstanding tax advice at lower cost from many nonattorney tax professionals such as enrolled agents and certified public accountants. These tax professionals can help you research IRS rulings and answer other tax-related questions.

Enrolled Agents

Enrolled agents are tax advisors and preparers licensed by the IRS. They earn the designation of enrolled agent either by passing a difficult IRS test or by working for the IRS for at least five years. An enrolled agent is generally the least expensive of the tax pros and is adequate for most small business tax advice and reporting.

Accountants

Certified public accountants (CPAs) are licensed and regulated by each state, like attorneys. They perform sophisticated accounting and business-related tax work and prepare tax returns. CPAs shine in giving business tax advice but generally are not as aggressive as tax lawyers when facing IRS personnel. Some states license accountants other than CPAs, such as public accountants. Many of these workers may also be able to give you competent advice on the tax effects of using ICs in your business.

Insurance Brokers

Insurance brokers and agents who sell workers' compensation insurance and business liability insurance should be knowledgeable about your state's workers' compensation laws. They can also help you save money when you buy such insurance.

Industry and Trade Associations

Business or industry trade associations or similar organizations can be a great source of information on IC issues. Many such groups track federal and state laws, lobby Congress and state legislatures, and even help members fight the IRS and other federal and state agencies. Find out if there is such an organization for your business or industry and get in touch with it.

Doing Your Own Legal Research

If you decide to investigate the law on your own, your first step should be to obtain a good guide to help you understand legal citations, use the law library or online research tools, and understand what you find. There are a number of sources that provide a good introduction to legal research, including *Legal Research: How to Find & Understand the Law,* by Stephen Elias and the Editors of Nolo (Nolo). This nontechnical book explains how to use all major legal research tools and helps you frame your research questions.

You might next want to find a law library. Your county should have a public law library, often at the county courthouse. Public law schools often contain especially good collections and generally permit the public to use their libraries. Some private law schools grant access to their libraries— sometimes for a modest fee. The reference department of a major public or university library may have a fairly decent legal research collection. Finally, don't overlook the law library in your own lawyer's office. Many lawyers, on request, will share their books with their clients.

Researching Federal Tax Law

Many resources are available to augment and explain the tax information in this book: IRS publications, self-help tax preparation guides, textbooks, court decisions, and periodicals. Some are free, and others are reasonably priced. Tax publications for professionals are expensive, but are often available at public or law libraries.

IRS Booklets

The IRS publishes over 350 free booklets explaining the tax code, and many are clearly written and useful. These booklets, called IRS Publications, range from several pages to several hundred pages in length. The following IRS Publications contain useful information for businesses that hire ICs:

- Publication 15, Circular E, *Employer's Tax Guide*
- Publication 15-A, *Employer's Supplemental Tax Guide*
- Publication 51, Circular A, *Agricultural Employer's Tax Guide*
- Publication 334, *Tax Guide for Small Business*
- Publication 505, *Tax Withholding and Estimated Tax*, and
- Publication 926, *Household Employer's Tax Guide*.

You can also download these publications from the agency's website at www.irs.gov.

> **CAUTION**
>
> **Don't rely exclusively on the IRS.** IRS publications are useful to obtain information on IRS procedures and to get the IRS's view of the tax law. But keep in mind that they present only the IRS's interpretation of the law, which may be very one-sided and even contrary to court rulings. So don't rely exclusively on IRS publications for information.

The *Internal Revenue Manual*

The *Internal Revenue Manual*, or *IRM*, is a series of handbooks that serve as internal guides to IRS employees on points of tax law and procedure. The *IRM* tells IRS auditors or collectors how specific tax code provisions should be enforced. Section 4.23 of the *IRM* deals with employment tax examinations and provides useful guidance on how IRS auditors handle worker classification questions. It also explains how employment tax audits are conducted. It will be particularly helpful if you're handling an IRS audit yourself. The *Internal Revenue Manual* is available on the IRS website at www.irs.gov/irm/index.html.

Internal Revenue Code

All federal tax laws are in the Internal Revenue Code, or IRC, which is written by Congress and often referred to as the tax code. The IRC is found in Title 26 of the United States Code, abbreviated as U.S.C. Title simply refers to the place within the massive U.S.C. where the IRC is found. The IRC is divided up into sections, which are subdivided and resubdivided into more parts. A reference to IRC § 3121(d)(3)(C) means that this particular tax law is found in Title 26 of the U.S.C., the Internal Revenue Code, Section 3121, subsection (d), paragraph (3), subparagraph (C).

The tax code is extremely long, complex, and difficult to understand. Fortunately, you probably don't need to read it. Few sections of the IRC deal with ICs, and they have all been summarized in this book.

The Internal Revenue Code is available online at www.law.cornell.edu/uscode/text/26.

IRS Pronouncements on Tax Law

The IRS makes written statements of its position on various tax matters. The statements do not have the force of law, but guide IRS personnel and taxpayers as to how specific tax laws should be interpreted and applied. Over the years, the IRS has issued thousands of rulings on how workers in almost every conceivable occupation should be classified.

Reviewing IRS rulings can help you predict how the IRS is likely to classify a worker if you're audited. However, IRS rulings are not always consistent and may even conflict with each other. Even IRS auditors don't always follow IRS rulings.

But, if you can find a favorable ruling, it can supply a reasonable basis for classifying a worker as an IC. Such a ruling might enable you to qualify for safe harbor protection. (See Chapter 3 for more about safe harbor protection.) Unfortunately, the majority of IRS rulings find that the workers involved are employees, not ICs, and will be of little help.

The IRS makes its rulings and legal summaries known in a number of publications:

- **Revenue Rulings** (Rev. Rul.) are IRS written statements of how the tax law applies to a specific set of facts. These are published as general guidance to taxpayers. The rulings are reprinted and indexed by IRC section and subject matter. A revenue ruling usually contains a hypothetical set of facts, followed by an explanation of how the tax code applies to those facts. For example: Rev. Rul. 2014-41 refers to IRS Revenue Ruling number 41, issued in 2014.

- **IRS Letter Rulings** are IRS answers to specific written questions and hypothetical situations posed by taxpayers. There are thousands of IRS letter rulings dealing with how to classify workers. Letter rulings are published in the *Internal Revenue Cumulative Bulletin* and in private tax service publications found in larger public and law libraries. For example: Ltr. Rul. 20142012 refers to a ruling issued in the 20th week of 2014; it was the 12th letter ruling issued that week.

- **IRS Revenue Procedures** (Rev. Procs.) are another way the IRS tells taxpayers how to comply with certain tax provisions, although they are primarily written to guide tax professionals and preparers. Revenue procedures often explain when and how to report tax items. They are published in the *Internal Revenue Cumulative Bulletin*, found in larger public and law libraries and widely reprinted in professional tax publications. For example: Rev. Proc. 2014-15 refers to a published revenue procedure numbered 15, issued in 2014.

- **IRS Regulations,** also called Treasury Regulations or Regs, are the IRS's most authoritative statement on how to interpret the IRC Regulations. They are usually found in a four-volume set called *Treasury Regulations*, found in most larger libraries and some bookstores. IRS Regulations are somewhat easier to read and comprehend than the tax code. However, few Regs concern IC issues.

Tax Cases

Federal courts have interpreted the tax law in thousands of court cases, many of which involve worker classification issues. A favorable court decision can also provide you with a reasonable basis for obtaining IRS saf e harbor protection. (See Chapter 3 for more about safe harbor protection.) You may rely on decisions by any federal tax court, federal district court, federal court of appeals, or the U.S. Supreme Court that haven't been overruled.

Directories of IRS Rulings

Finding an IRS ruling dealing with a situation similar to yours can be very difficult. Fortunately, tax experts have culled through thousands of IRS rulings on worker classification and categorized them by occupation. You can find them in a book called *Employment Status—Employee v. Independent Contractor* (391-3rd), published by BNA (www.bnatax.com). This book is expensive, so you may want to look for a copy in your local law library. Hundreds of IC rulings are also available on a subscription website at www. workerstatus.com.

In addition, many state chambers of commerce publish guides for companies that hire ICs, that list IRS rulings for various occupations. Call your state chamber of commerce to see if it publishes such a guide for your state; the chamber probably has an office in your state capitol.

Finally, an industry trade group or association may also be aware of, or even have copies of, helpful IRS rulings and court decisions.

Trade Association Publications

Every business or trade has its own publications and newsletters that closely track tax issues of common interest, including worker classification issues. By reading them, you can learn about recent IRS and court rulings affecting other employers in your industry.

Researching Other Areas of Law

Many fields of law other than federal tax law are involved when you hire ICs. These include other federal laws that apply throughout the country and the laws of your particular state.

Federal Laws

The best starting point for further research into federal employment and antidiscrimination laws that affect your business is *The Essential Guide to Federal Employment Laws,* by Lisa Guerin and Amy DelPo (Nolo). This guide will likely answer your questions, or at least tell you where to go for more information.

In addition, the U.S. Labor Department publishes a free pamphlet on worker classification for purposes of the federal labor laws called *Fact Sheet #13: Employment Relationship Under the Fair Labor Standards Act (FLSA)*. You can download a copy from the Labor Department's website at www. dol.gov.

State Laws

If you have questions about your state workers' compensation, unemployment compensation, tax, or employment laws, first contact the appropriate state agency for more information. Many of these agencies publish useful information pamphlets. This book contains contact information for state workers' compensation agencies (see Chapter 6), unemployment compensation agencies (see Chapter 5), tax departments (see Chapter 5), and labor departments (see Chapter 8).

More in-depth research into your state law will require that you review:

- legislation, also called statutes, passed by your state legislature (your state unemployment compensation, workers' compensation, and income tax withholding laws will probably be most important to you)
- administrative rules and regulations issued by state administrative agencies such as your state unemployment compensation agency, and
- published decisions of your state courts.

Many states, particularly larger ones, have legal encyclopedias or treatises that organize the state case law and some statutes into narrative statements organized alphabetically by subject. Through citation footnotes, you can locate the full text of the cases and statutes. These works are a good starting point for in-depth state law research.

Online Resources

A vast array of information for small business owners is available on the Internet.

Nolo's website, www.nolo.com, contains much useful information for businesspeople, including articles about independent contractors.

Somewhat surprisingly, the IRS has perhaps the most useful and colorful Internet site of any government agency. It contains virtually every IRS publication and tax form, and IRS announcements. The address is www. irs.gov.

How to Use the Interactive Forms

Editing RTFs..346

List of Forms...347

You can open, edit, save, and print the RTF files provided by this book using most word processing programs, such as Microsoft *Word*, Windows *WordPad*, and recent versions of *WordPerfect*. The forms in this book are available at: **www.nolo.com/back-of-book/HICI.html**

> **TIP**
> **Note to Macintosh Users.** These forms were designed for use with Windows. They should also work on Macintosh computers; however, Nolo cannot provide technical support for non-Windows users.

> **TIP**
> **Putting the clauses together.** Each of the clauses discussed in this book is available for download as its own file—you may choose to copy the clauses you've decided to use and paste them into a single document. These clauses are also grouped together to form a single IC Agreement template, in turn available as its own file.

Editing RTFs

Here are some general instructions about editing RTF forms in your word processing program. Refer to the book's instructions and sample agreements for help about what should go in each blank:

- **Underlines** indicate where to enter information. After filling in the needed text, delete the underline. In most word processing programs, you can do this by highlighting the underlined portion and typing CTRL-U.
- **Bracketed and italicized text** indicates instructions. Be sure to remove all instructional text before you finalize your document.
- **Optional text** gives you the choice to include or exclude text. Delete any optional text you don't want to use. Renumber numbered items, if necessary.
- **Alternative text** gives you the choice between two or more text options. Delete those options you don't want to use. Renumber numbered items, if necessary.

- **Signature lines** should appear on a page with at least some text from the document itself.

Every word processing program uses different commands to open, format, save, and print documents, so refer to your software's help documents for help using your program. Nolo cannot provide technical support for questions about how to use your computer or your software.

CAUTION

In accordance with U.S. copyright laws, the forms provided by this book are for your personal use only.

List of Forms

The following files are included in rich text format (RTF) and are available for download at: **www.nolo.com/back-of-book/HICI.html**

Form or Clause	File Name
Title of Agreement	Title.rtf
Names of IC and Hiring Firm	Names.rtf
Term of Agreement	AgreementTerms.rtf
Services to Be Performed	Services.rtf
Payment	Payment.rtf
Terms of Payment	PaymentTerms.rtf
Expenses	Expenses.rtf
Independent Contractor Status	Status.rtf
Business Permits, Certificates, and Licenses	Permit.rtf
State and Federal Taxes	Taxes.rtf
Fringe Benefits	FringeBenefits.rtf
Workers' Compensation	WorkersComp.rtf
Unemployment Compensation	Unemployment.rtf
Insurance	Insurance.rtf
Terminating the Agreement	Terminate.rtf
Exclusive Agreement	Exclusive.rtf

Form or Clause	File Name
Severability	Severability.rtf
Applicable Law	ApplicableLaw.rtf
Notices	Notices.rtf
No Partnership	NoPartnership.rtf
Assignment	Assignment.rtf
Resolving Disputes	ResolveDispute.rtf
Signatures	Signatures.rtf
Modifying the Agreement	Modify.rtf
Work at Your Premises	Premises.rtf
Intellectual Property Ownership	IPOwnership.rtf
Confidentiality	Confidentiality.rtf
Nonsolicitation	Nonsolicitation.rtf
Independent Contractor Questionnaire	Questionnaire.rtf
Documentation Checklist	Documentation.rtf
Worker Classification Checklist	WorkerClass.rtf
Independent Contractor Agreement	GeneralContractor.rtf

Contractor Screening Documents

Independent Contractor Questionnaire 351

Documentation Checklist 353

Worker Classification Checklist 355

Independent Contractor Questionnaire

Name: _____

Fictitious business name (if any): _____

Business address: _____

Business phone and fax: _____

Employer identification number or Social Security number: _____

Form of business entity (check one): ☐ Corporation   ☐ Sole Proprietorship
                                      ☐ Partnership   ☐ Limited Liability Company

1. Provide the name, address, and dates of service of all companies for which you have performed services as an independent contractor for the past two years. But please do not provide any information you have a duty to keep confidential.

2. Have you ever hired employees? If yes, please complete the following for each employee:

   Name: _____

   Address: _____

   Title: _____

   Salary: _____

   Dates of employment: _____

   Workers' compensation carrier and policy number: _____

3. Have you paid federal and state payroll taxes for your employees?   ☐ Yes ☐ No

4. Do you hold a professional license? If so, please provide a copy.

5. Do you have a business license? If so, please provide a copy.

6. Describe the training you have received in your specialty.

   School attended: _____

   Dates of attendance: _____ Degrees received: _____

   School attended: _____

   Dates of attendance: _____ Degrees received: _____

   School attended: _____

   Dates of attendance: _____ Degrees received: _____

7. Do you advertise your services?   ☐ Yes ☐ No

   If so, please provide a copy of these advertisements, including a yellow pages listing.

8. If you don't advertise, how do you market your services? _____
_____

9. Do you have a white pages business phone listing? ☐ Yes ☐ No
If so, please provide a copy.

10. Do you have a website marketing your services? If yes, please provide a copy of the home page and list the URL here: _____

11. Describe the business expenses you have paid in the past, including office or workplace rental, materials and equipment, telephone, and other expenses:
_____
_____
_____

12. Describe the business expenses you pay now: _____
_____
_____

13. Describe the equipment and facilities you own: _____
_____
_____

14. Describe the tools and materials you will use to perform the services in this job: _____
_____
_____

How much do they cost?_____

15. Please list your general liability insurance carrier: _____
_____ Policy number: _____

16. Please list your auto insurance carrier: _____
_____ Policy number: _____

17. Have you ever worked for us before? If so, please complete the following:
Dates of employment: _____
Services performed: _____

18. Do you have an independent contractor agreement? If so, please attach a copy.

19. Do you have your own business cards, stationery, and invoice forms? If so, provide copies.

20. If you're a sole practitioner, have you paid self-employment taxes on your income and filed a Schedule C with your federal tax return? ☐ Yes ☐ No

If so, will you provide copies of your tax returns for the past two years? ☐ Yes ☐ No

Documentation Checklist

Please provide the following documentation:

- ☐ copies of your business license and any professional licenses you have

- ☐ certificates showing that you have insurance, including general liability insurance and workers' compensation insurance if you have employees

- ☐ your business cards and stationery

- ☐ the URL for your website, if any, and a screenshot of the first page

- ☐ copies of any advertising you've done, such as a yellow pages listing

- ☐ a copy of your white pages business phone listing, if there is one

- ☐ if you're operating under an assumed name, a copy of the fictitious business name statement

- ☐ a copy of your invoice form to be used for billing purposes

- ☐ a copy of any office lease and proof that you've paid the rent, such as copies of canceled rental checks

- ☐ a photograph of your office or workplace

- ☐ your state unemployment insurance number

- ☐ copies of 1099 forms you received from other companies

- ☐ the names and salaries of all assistants that you will use on the job

- ☐ the names and salaries of all assistants you have used on previous jobs for the past two years and proof that you paid them, such as copies of canceled checks or copies of payroll tax forms

- ☐ a list of all the equipment and materials you will use in performing the services and how much they cost; proof that you have paid for the equipment, such as copies of canceled checks, is very helpful

- ☐ the names and addresses of other clients or customers for whom you have performed services during the previous two years; but don't provide any information you have a duty to keep confidential

Worker Classification Checklist

1. Is the worker a statutory independent contractor?

	Yes	No
Is the worker a direct seller?		
Is the worker a licensed real estate agent?		

If you answer "yes" to either question, treat the worker as an IC for federal tax purposes (but the worker's status for state unemployment and workers' compensation purposes must be determined under state tests, below).

2. Is the worker a statutory employee?

	Yes	No
Is the worker a corporate officer who provides services to the corporation?		
Is the worker a food, beverage, or laundry driver?		
Is the worker a traveling or city salesperson who sells goods for resale?		
Is the worker a full-time life insurance salesperson?		
Is the worker an at-home worker supplied with materials and specifications for work?		

If you answer "yes" to any question, treat the worker as an employee for federal tax purposes.

3. Is the worker an IC under the IRS common law test?

Complete the following chart. The more "yes" answers, the more likely the worker is an IC.

IRS Factors	Yes (supports IC status)	No (supports employee status)
1. **Instructions.** Will the firm **not** have the right to give the worker instructions about when, where, and how he or she is to do the job?		If NO, stop here. The worker is an employee.

IRS Factors	Yes (supports IC status)	No (supports employee status)
2. **Training.** Will the firm **not** give the worker training?		
3. **Significant investment.** Has the worker invested in facilities (such as an office) used to perform services?		
4. **Payment of expenses.** Will the worker be required to pay his or her own business or travel expenses?		
5. **Services available.** Does the worker make his or her services available to other businesses?		
6. **Method of payment.** Will the worker be paid by commission or by the job rather than by the hour, week, or month?		
7. **Realization of profit or loss.** Will the arrangement enable the worker to realize a profit or loss?		
8. **Written contract.** Will a written contract be executed describing the worker as an independent contractor?		
9. **Employee benefits.** Will the worker **not** receive any employee benefits?		
10. **Right to terminate.** Does the firm lack the legal right to terminate the worker at any time without incurring liability?		
11. **Regular business activity.** Is the work to be performed outside of the firm's regular business?		

4. Does the worker qualify for safe harbor protection?

	Yes	No
All 1099s filed		
Consistent treatment		
Reasonable basis for IC classification		

If all three boxes are checked "yes," treat the worker as an IC for federal employment tax purposes.

5. Must the worker be provided with state unemployment insurance coverage?

	Yes	No
Are the worker's pay or hours of work less than the state threshold requirements for UC coverage?		
Does an IC exemption apply—for example: • casual labor • spouse, child, or parent • commissioned real estate brokers and salespeople		
Is the worker an IC under state UC laws?		

If you answer "yes" to any question, you need not provide state unemployment insurance coverage.

6. Is the worker an IC for workers' compensation purposes?

	Yes	No
Does the employee minimum exclusion apply?		
Does the casual labor exclusion apply?		
Does the domestic worker exclusion apply?		
Does another exclusion apply?		
Is the worker an IC under state worker's compensation laws?		

If you answer "yes" to any question, you need not provide workers' compensation coverage.

Index

A

ABC test, 19
  behavioral control factors, 137, 138
  independent business or trade, 138, 141
  modified, 139, 141–142
  outside services factors, 137, 138, 140–141
  states that use, 139, 145, 146, 147
  for unemployment compensation
    purposes, 137–142
ACA. *See* Obamacare (Affordable Care Act)
Accountants, 336
  for appeals of IRS rulings, 109
  for classification issues, 88, 90–91
  incorporation of, 251
  instructions to, 27
  payment methods, 23
  for retirement plan issues, 128
Administrative appeals, to IRS, 108
Administrators and executives, FLSA
  exemption for, 218
  See also Corporate officers
Affordable Care Act. *See* Obamacare
Age Discrimination in Employment Act,
  219, 225
Agreements
  to accept CSP offers, 112
  to avoid statutory employee
    classification, 74
  with leasing companies, 259
  real estate documents, 284
  sale of goods contracts, 284

  See also Independent contractor
    agreements
Agricultural workers, 144, 164, 219
Aliens. *See* Undocumented workers
Americans with Disabilities Act, 9, 219,
  226, 227
Antidiscrimination laws
  Age Discrimination in Employment Act,
    219, 225
  Americans with Disabilities Act, 9, 219,
    226, 227
  Civil Rights Act, 9, 219, 225, 228
  employee lawsuits and, 6
  Equal Pay Act, 226
  federal, 225–228
  Immigration Reform and Control Act
    and, 226
  leased workers and, 255
  local laws, 228
  Pregnancy Discrimination Act, 226
  researching, 341
  retirement plans and, 127
Appealing IRS rulings, 107–109, 124
Applicable law and forum, IC agreement
  provisions, 309
Approval requirement, IC agreement
  provisions, 312–313
Arbitration, IC agreement provisions,
  314–315
Architects, classification rulings, 62
Articles of incorporation, 248, 251

"Assigned risk pools," workers' compensation insurance, 181–182

Assignments and assignment agreements for copyrights, 239–240, 284
IC agreement provisions, 311–312, 321–322
for trade secrets/patents, 241–242

Assistants. *See* Employees, of independent contractors

Athletes (professional), workers' compensation for, 165

Attachments, to IC agreements, 291

Attorneys. *See* Lawyers

At-will employment
right to fire, 7, 22, 57, 168, 170
right to quit, 25, 57

Audits by IRS, 42, 101–128
20-factor test, 114
actions that do not constitute, 94
appealing, 107–109, 124
auditor procedures, 104–105
basics, 8–9, 101–102
examination reports, 107
incorporated workers and, 249–251
leased workers and, 253
legal advice, 335
of retirement plans, 97, 126–128
right-of-control evaluations, 20–21, 128
safe harbor based on prior audits, 92–94
sole proprietorships and, 245
standard-form IC agreements and, 285
statute of limitations, 103–104
what you need to prove, 105–106
who gets audited, 102–103
See also Classification, IRS rules;
Classification Settlement Program
(CSP); Classification tests; Penalties for
worker misclassification, federal

Audits by labor-related federal agencies, 8, 113

Audits by states, 10–11
information shared with IRS, 103, 132
standard-form IC agreements and, 285
unemployment compensation audits, 10, 103, 131–132
workers' compensation audits, 10, 130, 132, 168, 179, 248

B

Babysitters. *See* Household workers

Backup withholding, for prospective workers, 265–267

Barbershop workers, 63–66

Beauty salon workers, 63–66, 84–85

Behavioral control
ABC test, 137, 138
common law test and, 49–53, 66, 128, 167, 170, 223, 227
loss of, as drawback of ICs, 11

Benefits. *See* Employee benefits

Beverage distributors. *See* Drivers

Breach of contract
of IC agreements, 307
wrongful termination suits, 7, 12, 168

Broadcast workers, FLSA exemption, 219

Business expenses
common law test and, 25, 50, 54–55, 70
IC agreement provisions, 299–300
reimbursing ICs for, 269–271
travel and entertainment expenses, 25, 50, 70

Business permits, IC agreement provisions, 302

C

California
  copyrights for commissioned works, 237
  overtime compensation, 195
  unemployment coverage rules, 145
  Uninsured Employers Benefits Trust
    Fund, 158
  workers' compensation classification,
    173–174, 181
Camp employees, FLSA exemption for, 219
Caretakers. *See* Companion sitters;
  Household workers
Casual labor
  defined, 149
  FLSA exemption for, 219
  household workers as, 196
  temporary workers, 13, 149
  unemployment coverage for, 149
  workers' compensation for, 161–162
  See also Part-time workers
Certificates
  IC agreement provisions, 302
  required for prospective workers, 263
Certified public accountants (CPAs), 336
Chauffeurs. *See* Drivers; Household
  workers
Children. *See* Family members as workers;
  Minor children
Children (adult), parents employed by, 200
Church employees and clergy
  FLSA exemption for, 219
  workers' compensation for, 165
Citizenship status
  green cards, 43, 197
  verification by employers, 230
  See also Undocumented workers
Civil Rights Act, 9, 219, 225, 228

Classification
  by non-IRS agencies, 95
  of prospective workers, 264
Classification for unemployment
  compensation, 134–150
  ABC test, 137–138, 140–141
  for agricultural workers, 144
  basics, 134–135
  common law test, 136
  for construction industry, 145–147
  for domestic workers, 144
  list of state tests, 139, 142
  modified ABC tests, 141–142
  in Montana, 143
  for statutory employees, 147–148
  in Wisconsin, 143
  in Wyoming, 143–144
Classification for workers' compensation,
  165–177
  basics, 165–166
  in California, 173–174, 181
  certification of IC status, 167
  common law test, 167–170
  consequences of misclassifying, 177
  for construction industry, 176, 181
  in Michigan, 174
  in Minnesota, 174–175
  in Washington, 175–176
  in Wisconsin, 176
Classification, IRS rules, 1–2, 8–9, 40–97
  asking IRS for determination, 44–45
  classification rulings, 94
  copyright ownership, 240–241
  flow chart, 44
  incorporated workers and, 249–251
  rulings for specific occupations, 61–71
  Step 1, 40, 46–48
  Step 2, 49–71

Step 3, 42, 72–80
Step 4, 42, 80–97
Worker Classification Checklist, 355–357
worker misclassification lawsuits, 11
for workers' compensation purposes, 165–166
See also Audits by IRS; Common law test, IRS use of; Penalties for worker misclassification, federal; Safe harbor protection
Classification Settlement Program (CSP), 107, 109–115
basics, 109–110
offer types, 110–111
procedures, 112
pros and cons of accepting, 112, 114–115
voluntary (VCSP), 113–115
Classification tests, list of state tests, 139, 142, 169
See also ABC test; Common law test; Economic reality test; Relative nature of work test
Clients and customers
multiple clients, 23–24, 26, 29, 57, 171
one client, 57
Commissions, 13
safe harbor protection for commissioned workers, 85
statutory independent contractors and, 46
unemployment coverage and, 148
Common law test, 17–37
control-measurement factors, 20–33, 167–168, 170
copyright ownership and, 240–241
Employment Determination Guide, 33–37
non-IRS agency use of, 95
right of control basics, 19–20
states that use, 139, 169, 173–174

for unemployment compensation purposes, 136
when required, 17–18
for workers' compensation purposes, 167–170
Common law test, IRS use of, 18, 42, 49–71
applying the test, 59–61
for audit appeals, 107
audits and, 106
behavioral control factors, 49–53, 66, 128, 223, 227
examples, 59–61
financial control factors, 49–50, 53–56
for immigration purposes, 230
misclassification and, 116–117
for Obamacare purposes, 211
relationship of parties factors, 49–50, 56–59
rulings for specific occupations, 61–71
Companion sitters
classification rulings, 71
overtime/minimum wage exemption, 196
Computer consultants and programmers
equipment used by, 23
FLSA exemption for, 218–219
instructions to, 26, 27
safe harbor protections, 96–97
statutory employee rules, 77
Confidentiality
IC agreement provisions, 324–325
of trade secrets, 242
Consistent treatment requirement, safe harbor protection, 81, 83–87, 88, 105, 113
Construction workers
unemployment coverage for, 145–147
workers' compensation requirements, 165, 176, 181
Consultants, statutory employee rules, 76, 77

See also Computer consultants and programmers

Contacts adjustments, IRS, 94

Continuing relationship
common law test and, 24
with employees, 11
with statutory employees, 74

Contracts. *See* Agreements; Independent contractor agreements

Control over workers. *See* Behavioral control; Common law test

Copyrights, 12, 232–233
assignment of, 239–240, 284
coauthors, 235
common law test and, 18
employee classification concerns, 240–241
failure to obtain transfer of, 235
on works made for hire, 234, 235–238
on works not made for hire, 234, 238–240

Corporate officers, 248–249
personal guarantees, 284
statutory employee rules for, 72, 74–75, 147
workers' compensation for, 164

Corporations, 248–252
1099 filing requirement, 274
antidiscrimination laws, 227
family members employed by, 198, 199
finding incorporated ICs, 251–252
IC agreement signatures, 290
incorporation, 248, 251, 252
IRS audits of, 104
pros and cons of hiring incorporated workers, 249–251
shareholders, 211, 248–249
taxpayer ID numbers, 317
workers' compensation and, 177

Court litigation. *See* Lawsuits; Lawsuits by employees

CPAs (certified public accountants), 336

Credit reports, FCRA protections, 224–225

Criminal sanctions
failure to provide workers' comp, 177
for tax fraud, 125

CSP. *See* Classification Settlement Program

Customers and clients
multiple clients, 23–24, 26, 29, 57, 171
one client, 57

Customs in trade or industry, common law test and, 32–33

D

Department of Health and Human Services (HHS), 9, 206

Department of Labor. *See* U.S. Department of Labor

Department of Treasury (U.S.), misclassification audits by, 9

Designers, safe harbor protections, 96–97

Directors, corporate, 248–249
See also Corporate officers

Direct sellers
FLSA exemption for, 218–219
as statutory independent contractors, 40, 46–47, 211

Disability insurance/taxes, state, 1, 6, 150, 152

Disabled workers, Americans with Disabilities Act, 9, 219, 226, 227

Discrimination. *See* Antidiscrimination laws

Dispute resolution, IC agreement provisions, 313–316

Doctors
  incorporation of, 251
  instructions to, 27
Documentation
  audit role, 91, 94, 105–106
  for FLSA requirements, 222
  IC file requirements, 267
  requirements for prospective workers,
    263–264
  for travel/entertainment expenses
    incurred by ICs, 270–271
Documentation Checklist, 353
Domestic workers. *See* Household workers
Door-to-door salespeople. *See* Direct sellers
Draftspeople, classification rulings, 62
Drivers
  FLSA exemption for, 219
  limousine, 66–67
  safe harbor protection, 85
  statutory employee rules for, 72,
    77–78, 147
  taxi, 68
  truck, 69–71
  van operators, 67
  workers' compensation for, 165

E

Economic reality test, 19
  for antidiscrimination statutes, 227
  in California, 173–174
  FLSA use of, 193, 219–222
EEOC (Equal Employment Opportunity
  Commission), 9, 220
EINs (employer identification numbers),
  190–191, 266, 317
Elder care, companion sitters, 71, 196
  See also Household workers
Electronic signatures, 287, 318

Emailing IC agreements, 287, 291–292
Employee benefits, 4–6
  common law test and, 31, 50, 58
  IC agreement provisions, 303–304
  sick leave, 6, 31
  vacation leave, 6, 31
  worker lawsuits to contest, 128
  See also Health insurance; Retirement
    plans
Employee Retirement Income Security Act
  (ERISA), 9, 126–128
Employees
  complaints to IRS by, 102
  defined for workers' comp purposes, 167
  disadvantages of using, 4–8
  dual-status workers, 49, 87
  full-time equivalent, 207–209, 212–214
  household workers as, 192
  leasing, 211, 252–259
  liability for actions of, 7
  right to fire, 7, 22, 57, 168, 170
  right to quit, 25, 57
  short-term, 254
  telecommuting, 254
  trade secrets for inventions by, 241–242
  at-will employment, 25
  works made for hire by, 236, 240–241
  See also Classification, IRS rules;
    Classification tests; Lawsuits by
    employees; Part-time workers; Statutory
    employee rules
Employees, of independent contractors
  common law test and, 28
  documentation of, 264
  workers' compensation for, 178, 179–181
Employer identification numbers (EINs),
  190–191, 266, 317
Employer shared responsibility payment,
  Obamacare, 211–215

Employment contracts, 22

Employment Determination Guide, 33–37

Employment laws, information
resources, 205

Employment Plans and Exempt
Organizations Divisions (EP/EO), 97

*Employment Status--Employee v.
Independent Contractor*, 341

Employment Tax Examination (ETE), 103

*Employment Tax National Research
Project*, 102

Engineers, 27, 96–97

Enrolled agents, 336

Entertainment expenses. *See* Travel and
entertainment expenses

Equal Employment Opportunity
Commission (EEOC), 9, 220

Equal Pay Act, 226

Equipment and facilities, investment in
common law test and, 24–25, 50, 54,
67, 69–70
economic reality test and, 220
statutory employee rules, 73, 74
See also Tools and materials

ERISA (Employee Retirement Income
Security Act), 9, 126–128

Estates, family members employed by,
198, 199

Estimated taxes, for household workers,
189–190

Examination reports, IRS audits, 107

Exclusive agreement, IC agreement
provisions, 308

Executives and administrators, FLSA
exemption for, 218
See also Corporate officers

Expenses. *See* Business expenses

Experience and skill requirements,
common law test and, 31–32

F

Facilities. *See* Equipment and facilities,
investment in

*Fact Sheet #13: Employment Relationship
Under the Fair Labor Standards Act*, 342

Fair Credit Reporting Act (FCRA),
224–225

Fair Labor Standards Act (FLSA), 215–222
avoiding problems, 222
basics, 215–216
covered businesses, 217
employee classification rules, 9, 193,
219–222
protections for household workers,
192–196
record-keeping requirements, 222
workers exempt from overtime/
minimum wage requirements, 217–219

Family and Medical Leave Act (FMLA),
9, 223–224

Family leave insurance, 152

Family members as workers, 198–201
federal payroll taxes, 198–200
state payroll taxes, 201
unemployment coverage for, 148
workers' compensation for, 165

Farmworkers. *See* Agricultural workers

Faxing IC agreements, 287, 291–292

Federal district court appeals, 109

Federal employer identification numbers
(EINs), 190–191, 266, 317

Federal laws
Age Discrimination in Employment Act,
219, 225
Americans with Disabilities Act, 9, 219,
226, 227
Civil Rights Act, 9, 219, 225, 228

Immigration Reform and Control
Act, 226
Internal Revenue Code, 80, 339
National Labor Relations Act, 9,
222–223
Occupational Safety and Health Act,
9, 229
Pregnancy Discrimination Act, 226
researching, 341–342
Social Security Act, 9
Worker Adjustment and Retraining
Notification Act, 219
See also Fair Labor Standards Act (FLSA);
Obamacare (Affordable Care Act)
Federal tax courts, 340–341
FedEx delivery drivers, 67
FICA (Social Security and Medicare
taxes), 1
basics, 4–5, 41
for children employed by parents,
198–199
failure to pay, 96
for household workers, 185, 186–190
for incorporated workers, 248–249
for leased workers, 253, 259
for one spouse employed by another,
199–200
for parents employed by children, 200
for partnerships, 246
payment offsets, 122
penalties for misclassification and,
117–121
safe harbor protection, 81
Social Security tax ceiling, 118
for sole proprietorships, 245
for statutory independent contractors,
46, 47, 48
Fictitious business names, 263

Financial control, common law test,
49–50, 53–56
Firing. *See* Right to fire
Fishing boat crews, as statutory
independent contractors, 48
FLSA. *See* Fair Labor Standards Act
FMLA (Family and Medical Leave Act),
9, 223–224
Food distributors. *See* Drivers
Forms
editing, 346–347
independent contractor agreements,
283, 325
list of, 347–348
See also specific IRS forms
Fraud rule, 104
Fraudulent actions
criminal sanctions for, 125, 177
penalties for, 10, 125, 130, 314
safe harbor protection and, 92
state laws, 10, 284
See also Penalties for worker
misclassification, federal; Penalties for
worker misclassification, state
Fringe benefits. *See* Employee benefits
Full-time equivalent employees for ACA,
207–209, 212–214
FUTA (federal unemployment tax), 1
for agricultural workers, 144
basics, 41
for children employed by parents,
198–199
common law test and, 136
for household workers, 187, 188–190
for incorporated workers, 248–249
for leased workers, 253, 259
for one spouse employed by another,
199–200

for parents employed by children, 200

for partnerships, 246

penalties for misclassification and, 117–121

pre-2011 tax rate, 120

safe harbor protection, 81

for sole proprietorships, 245

for statutory independent contractors, 46, 47, 48

tax rate, 131

tax thresholds, 133–134

G

Gardeners. *See* Household workers

Government (non-IRS) agencies, classification determinations by, 95
See also State agencies; *specific governmental agencies*

Green cards (immigrant visas), 43, 197

H

Health aids. *See* Household workers

Health and Human Services Department (HHS), 9, 206

Health insurance
common law test and, 31, 50, 58
IC agreement provisions, 303–304
See also Obamacare (Affordable Care Act); Workers' compensation insurance

Hiring
by ICs, common law test and, 28
non-U.S. citizens, 43
undocumented workers, 6, 197, 230

Homeowner's insurance, for injured household workers, 191–192

Hours of work. *See* Working hours

Household workers, 184–197
from agencies, 186
children employed by parents as, 198–199
companion sitters, 71, 196
EINs for, 190–191
federal payroll tax status, 184–191
FLSA exemption for, 219
independent contractor agreements, 186
insurance for injuries to, 191–192
live-in, 196, 219
minimum wage laws, 192–196
overtime compensation, 192–193, 195–196
safe harbor protection, 185
statutory employee rules for, 72, 75–77
undocumented, 197
unemployment coverage for, 144
workers' compensation for, 162–163

I

ICs. *See* Independent contractors

"If any" workers' compensation policies, 178

Illegal aliens. *See* Undocumented workers

Illinois, unemployment coverage rules, 145

Immigrant visas (green cards), 43, 197

Immigration laws, 226, 230

Immigration Reform and Control Act, 226

Income taxes, failure to pay, 103

Income tax withholding, federal, 1, 4–5, 41
for children employed by parents, 199
failure to withhold, 96
for household workers, 185, 187
IC agreement provisions, 302–303, 317
for incorporated workers, 248–249
leased workers and, 253, 259

nonresident alien exemption, 43
for parents employed by children, 200
payment offsets, 121–122
penalties for misclassification and,
    117–121, 124
safe harbor protection, 81, 83–84, 87
for sole proprietorships, 245
Income tax withholding, state, 1, 153
audits related to, 132
failure to withhold, 96
for family members as workers, 201
IC agreement provisions, 302–303
misclassification audits, 10
reporting on federal forms, 87
Incorporation, 248, 251, 252
See also Corporations
Indemnification
IC agreement provisions, 321–322
by leasing companies, 259
See also Penalties for worker
    misclassification, federal; Penalties for
    worker misclassification, state
Independent business or trade
ABC test, 138, 141
relative-nature-of-the-work test, 171
Independent contractor agreements
attachments and exhibits, 291
common law test and, 50, 58
copies of, 291
dates, 290–291
downloading, 293, 326–331
drafting, 285–287
electronic contracts, 286–287
as evidence of IC status, 13, 46, 284–285
faxing or emailing, 287, 291–292
for household workers, 186
modifying, 291, 318–320
obeying terms of, 269

oral vs. written, 283–284
for prospective workers, 265
provided by ICs vs. using own, 265, 286
samples, 283, 326–331
signatures, 287, 288–290, 317–318
standard-form agreements, 285
terminating, 7, 12, 307
Independent contractor agreements,
    essential provisions
applicable law and forum, 309
approval requirement, 312–313
assignment, 311–312
court litigation, 316
dispute resolution, 313–316
exclusive agreement, 308
expenses, 299–300
fringe benefits, 303–304
independent contractor status, 300–302
liability insurance, 306
names of parties, 293
no partnership, 310–311
notices, 309–310
payment, 295–297
permits, certificates, and licenses, 302
services to be performed, 294–295
severability, 308
signatures, 287, 288–290, 317–318
termination of agreement, 307
term of agreement, 294
terms of payment, 297–299
title of agreement, 292–293
unemployment compensation, 305–306
workers' compensation, 304–305
Independent contractor agreements,
    optional provisions
confidentiality, 324–325
confidentiality clauses, 242
indemnification, 321–322

intellectual property ownership, 321–323
modifying the agreement, 318–320
nonsolicitation, 325
working at your premises, 320
Independent Contractor Questionnaire, 351–352
Independent contractors (ICs)
ACA employment mandate, 209–210
basics, 1, 17
benefits of using, 1, 4–8
documents required before hiring, 263–264
dual-status workers, 49, 87
end-of-service documentation, 271–280
household workers as, 191–192
incorporated, 248–252
interviewing prospective, 262–263
license requirements, 7, 13
LLCs, 246–248
myths about hiring, 13
nonresident aliens, 43
partnerships, 246
reimbursing for expenses, 269–271
risks of using, 8–12
self-employment taxes, 103
sole proprietorships, 244–245
statistics, 1
statutory, 40, 46–48
termination restrictions, 7, 12
workers' compensation for, 157–158, 178–181
work habits to avoid, 267–269
works specially commissioned by, 236–237, 240–241
See also Employees, of independent contractors

Independent contractor status, 1
certification for workers' comp purposes, 167
documents to support, 263–264
IC agreement provisions, 300–302
IC agreements to establish, 13, 46, 284–285
See also Classification, IRS rules; Classification tests
Industry practices
common law test and, 32–33
safe harbor protection and, 88–89
Injuries
insurance for household workers, 1 91–192
liability for employee injuries, 7
liability for IC injuries, 12, 306
OSHA protections, 9, 229
See also Workers' compensation insurance
Instructions to workers, common law test and, 25–27, 50, 51–52
Insurance
homeowner's insurance, 191–192
liability insurance, 12, 159, 179, 306, 336–337
life insurance, 6
renter's insurance, 192
See also Health insurance; Workers' compensation insurance
Insurance brokers and salespeople, 336–337
statutory employee rules for, 72, 78–79
unemployment coverage for, 148
workers' compensation for, 165
Integration into business, common law test and, 29–30

Intellectual property, 232–242
  assignment provisions, 321–322
  copyrights, 12, 232–233, 234–241
  IC agreement provisions, 321–323
  patents, 233, 241–242
  trade secrets, 234, 241–242
  work-made-for-hire agreements, 322–323
Intent of hiring firm and worker, common
  law test and, 32
Interest assessments by IRS, 125
Internal Revenue Code (IRC), 80, 339
Internal Revenue Code Section 530. *See*
  Safe harbor protection
Internal Revenue Manual (IRM), 338
Internal Revenue Service (IRS)
  ACA no-coverage penalty website, 212
  contacts adjustments by, 94
  Employment Tax Examination, 103
  Fishing Tax Center website, 48
  Revenue Procedures, 340
  Treasury Regulations, 340
  website, 338, 343
  See also Audits by IRS; Classification,
    IRS rules; Classification Settlement
    Program (CSP); Common law test, IRS
    use of; *specific IRS forms*
Internal Revenue Service rulings
  appealing, 107–109, 124
  classification rulings for specific
    occupations, 61–71
  common law test and, 61–71
  Letter Rulings, 340
  No Change Letters, 107
  Revenue Rulings, 339–340
  safe harbor protection and, 88, 91–92
Inventions, trade secret/patent ownership,
  233, 234, 241–242

IRS Form 940, *Employer's Annual Federal
  Unemployment Tax Return*, 83, 87
IRS Form 941, *Employer's Quarterly
  Federal Tax Return*, 83
IRS Form 944, *Employer's Annual Federal
  Tax Return*, 103
IRS Form 1040, Schedule E, 246
IRS Form 1096, transmittal form, 275, 277
IRS Form 1099-MISC
  audit role, 105, 113, 271–272
  classification and issuance of, 13
  discrepancies, 94
  end-of-service filing of, 271–278
  filing extensions, 278
  filing procedures, 275–278
  household workers and, 186, 190
  incorporated workers and, 250, 274
  late filing, 82
  for LLCs, 247
  Matching Program, 103
  online filing, 278
  payments exempt from filing
    requirements, 273–274
  payments to lawyers, 275
  penalties related to, 117–119, 124
  safe harbor rules and, 81–83, 272
  services in course of trade or business,
    272–273
  statutory employee rules and, 75
  to support IC status, 264
  TINs and, 266
  when to file, 82–83, 272
IRS Form 4669, *Statement of Payments
  Received*, 121–122, 279
IRS Form 4670, *Request for Relief From
  Payment of Income Tax Withholding*,
  121–122, 279

IRS Form 8233, *Exemption From Withholding on Compensation for Independent (and Certain Dependent) Personal Services of a Nonresident Alien Individual*, 43

IRS Form 8809, *Application for Extension of Time To File Information Returns*, 278

IRS Form 8928, *Return of Certain Excise Taxes Under Chapter 43 of the Internal Revenue Code*, 212

IRS Form 8952, Application for *Voluntary Classification Settlement Program*, 113

IRS Form SS-4, *Application for Employer Identification Number*, 191

IRS Form SS-8, *Determination of Worker Status for Purposes of Federal Employment Taxes and Income Tax Withholding*, 44–45, 94

IRS Form W-2, *Wage and Tax Statement*, 58, 83, 103, 124, 185

IRS Form W-9, *Request for Taxpayer Identification Number and Certification*, 266

IRS Publication 15-A, *Employer's Supplemental Tax Guide*, 40, 338

IRS Publication 15, Circular E, *Employer's Tax Guide*, 40, 41, 65, 70, 338

IRS Publication 51, Circular A, *Agricultural Employer's Tax Guide*, 338

IRS Publication 334, *Tax Guide for Small Business*, 338

IRS Publication 505, *Tax Withholding and Estimated Tax*, 190, 338

IRS Publication 515, *Withholding of Tax on Nonresident Aliens and Foreign Entities*, 43

IRS Publication 560, *Retirement Plans for Small Business*, 258

IRS Publication 926, *Household Employer's Tax Guide*, 189, 338

IRS Publication 2108A, *On-Line Taxpayer Identification Number (TIM) Matching Program*, 266

IRS Schedule SE, 245

J

Joint employment, leased employees, 254–256

K

Key services, common law test and, 59

L

Labor Department, federal. *See* U.S. Department of Labor

Labor departments, state, 228

Labor laws, researching, 341
   See also Fair Labor Standards Act (FLSA); Federal laws; State laws

Laundry distributors. *See* Drivers

Lawsuits
   for IC injuries, 12
   oral agreements and, 283–284

Lawsuits by employees
   breach of contract, 7
   for discrimination, 6
   for health, safety, labor, antidiscrimination violations, 205
   for job-related injuries, 157–158, 177
   reduced exposure to, 6–7
   for worker misclassification, 11, 128
   for wrongful termination, 7

Lawyers, 334–335
  for appeals of IRS rulings, 109
  classification advice from, 75, 88, 90–91
  classification of, 27, 62–63
  finding, 335
  incorporation of, 251
  payment methods, 23
  reporting payments to, 275
  for retirement plan issues, 128
  tax attorneys, 335
Leased employees, 252–259
  basics, 252–253
  employment status, 253
  joint employment, 254–256
  retirement plans and, 257–258
  as statutory independent contractors, 211
  workers' compensation insurance and,
    253, 255, 256–257
  written agreements, 259
Leasing companies, 253, 255, 259
Legal research, 337–343
  federal laws, 337–342
  online resources, 342–343
  state laws, 342
Letter Rulings, IRS, 340
Liability
  for employee injuries, 157–158
  for IC injuries, 7, 12, 157–158, 306
  See also Lawsuits; Lawsuits by employees
Liability insurance
  buying, 336–337
  IC agreement provisions, 306
  for IC injuries, 12, 159, 179
Licenses, 13
  for barbers and beauticians, 64
  IC agreement provisions, 302
  liability for hiring unlicensed ICs, 7
  for real estate agents, 47

  required for prospective workers, 263
  for truck drivers, 70
Life insurance, employee benefit, 6
Life insurance salespeople, statutory
  employee rules for, 72, 78–79
Limited liability companies (LLCs), 164,
  247–248, 289
Limousine drivers, classification rulings,
  66–67
Litigation. See Lawsuits; Lawsuits by
  employees
Location of work
  for home workers, 75
  IC agreement provisions, 320
  on-site work, 22, 57
  telecommuting, 254
  UC laws and, 135–136

M

Maine, unemployment coverage rules, 146
Maryland, unemployment coverage
  rules, 146
Material breaches
  of IC agreements, 307
  wrongful termination suits, 12, 168
Materials. See Tools and materials
Mechanics, FLSA exemption for, 219
Mediation, IC agreement provisions, 313
Medicare taxes. See FICA (Social Security
  and Medicare taxes)
Michigan, workers' compensation
  classification, 174
Minimum wage
  FLSA requirements, 216
  for household workers, 192, 194–196
  laws mandating, 6
  room and board and, 195

workers exempt from FLSA requirements, 217–219

Minnesota
unemployment coverage rules, 146
workers' compensation classification, 174–175

Minor children, payroll taxes for children employed by parents, 198–199

Misclassification Initiative, FLSA, 216

Modifying IC agreements, 291, 318–320

Montana, UC classification test, 143

Motor vehicle salespeople, FLSA exemption for, 219

Movie theater employees, FLSA exemption for, 219

Multiple clients
common law test and, 23–24, 26, 29, 57
relative-nature-of-the-work test, 171

N

Nannies. *See* Household workers

National Labor Relations Act (NLRA), 9, 222–223

National Labor Relations Board (NLRB), 8, 9, 18, 222

Nebraska, unemployment coverage rules, 146

Newspaper sellers and distributors, 47, 165, 219

New York, unemployment coverage rules, 146–147

No Change Letters, IRS audits, 107

Nondisclosure, IC agreement provisions, 324

Nonprofit employees, 149, 165

Nonresident aliens
antidiscrimination laws, 226
federal tax withholding exemption, 43
immigration assistance for, 197
verifying citizenship status of, 197, 230
See also Undocumented workers

Nonsolicitation, IC agreement provisions, 325

No partnership, IC agreement provisions, 310–311

Notices, IC agreement provisions, 309–310

Nurses, private. *See* Household workers

O

Obamacare (Affordable Care Act), 206–215
basics, 5–6, 206
calculating number of employees, 207–209, 213, 214
employer mandate, 205, 209–210
for independent contractors, 209–210
"large employers" definition, 207
no-coverage penalties, 81, 112, 115, 126, 205, 211–215

Occupational Safety and Health Act (OSHA), 9, 229

Office space and equipment, for employees, 5

On-site work, common law test and, 22, 57

Outsourcing. *See* Leased employees

Overtime compensation
FLSA requirements, 216
for household workers, 192–193, 195–196
workers exempt from FLSA requirements, 217–219

P

Partnerships, 246
  construction industry rules, 145
  family members employed by, 198, 199
  IC agreement signatures, 288–289
  IRS audits of, 104
  partners as statutory ICs, 211
  taxpayer ID numbers, 317
  workers' compensation and, 164, 177
Part-time workers, 254
  classification of, 13, 59–60
  continuing relationship with, 74
  unemployment coverage for, 149
  workers' compensation for, 162
Patents, 233, 241–242
Payment methods
  for children employed by parents, 199
  common law test and, 23, 50, 56, 63,
    70, 168, 170
  IC agreement provisions, 295–299
  overtime compensation, 192–193,
    195–196, 216, 217–219
  See also Commissions
Payroll taxes, federal, 1, 4–5, 41
  audits of, 103
  for children employed by parents,
    198–199
  common law test and, 32, 33, 42
  failure to pay, 104, 114
  for household workers, 185, 186–190
  for ICs, 28
  for incorporated workers, 248–249
  IRS publications on, 40
  for one spouse employed by another,
    199–200
  paid by leasing companies, 253, 255, 259
  for parents employed by children, 200

  for partnerships, 246
  safe harbor protection, 81
  for sole proprietorships, 245
  for statutory employees, 72, 76, 78–79
  for statutory independent contractors,
    46, 47, 48
  See also FICA (Social Security and
    Medicare taxes); FUTA (federal
    unemployment tax); Income tax
    withholding, federal
Payroll taxes, state, 1, 46, 130
  common law test and, 32, 33
  disability insurance, 1, 6, 150, 152
  for family members as workers, 201
  for incorporated workers, 251
  paid by ICs, 28
  paid by leasing companies, 253, 259
  See also Income tax withholding, state;
    Unemployment compensation, state;
    Workers' compensation insurance
Penalties for worker misclassification,
  federal, 42
  ACA no-coverage penalties, 81, 112,
    115, 126, 205, 211–215
  appealing, 124
  basics, 116–117
  criminal sanctions, 125
  forms required to pay, 279
  IC agreement provisions, 321–322
  intentional misclassification, 116,
    120–122
  interest assessments, 125
  IRS Form 1099-MISC and, 117–119,
    124, 271–272
  legal/accountant assistance, 109
  legal implications table, 9–10
  miscellaneous penalties, 124–125

paid by leasing companies, 259
retirement plan audits, 126–128
trust fund recovery penalty, 123–124
unintentional misclassification, 116–120
See also Classification Settlement
Program (CSP)
Penalties for worker misclassification, state,
10–11, 130, 132, 177
Pennsylvania, unemployment coverage
rules, 147
Performance evaluations, common law test
and, 50, 53
Permits, IC agreement provisions, 302
Personal guarantees, by corporate
officers, 284
Personal performance of services
common law test and, 28
statutory employee rules, 73
Plant closings and mass layoffs, Worker
Adjustment and Retraining Notification
Act and, 219
Pregnancy Discrimination Act, 226
Professionals, FLSA exemption for, 218
Profit or loss opportunities, common law
test and, 21, 50, 56, 65, 66

Q

Quitting. *See* Right to quit

R

Rail, air, and motor carrier employees, 219
Real estate documents, written, 284
Real estate salespeople and agents
license requirements, 47
as statutory independent contractors,
40, 47–48, 211

unemployment coverage for, 148
workers' compensation for, 165
Recordkeeping. *See* Documentation
Reimbursing ICs, for travel and
entertainment expenses, 269–271
Relationship of parties, common law test,
49–50, 56–59
Relative-nature-of-the-work test, for
workers' compensation, 170–173
Relatives. *See* Family members as workers
Renter's insurance, 192
Reports (oral or written), common law test
and, 29
See also Documentation
Resident aliens, federal tax withholding
exemption, 43
Retirement plans
antidiscrimination rules, 127
common law test and, 31
for employees, 6
IC agreement provisions, 303–304
IRS audits of, 97, 126–128
leased workers and, 257–258
loss of tax-qualified status, 127–128
professional advice, 128
safe harbor protection, 97
tax-qualified plans, 126
worker lawsuits to contest, 128
Revenue Procedures, IRS, 340
Revenue Rulings, IRS, 339–340
Right to control test. *See* Common law test
Right to fire
common law test and, 22, 57, 168, 170
wrongful termination suits, 7
Right to quit, common law test and, 25, 57
Right to reject job offers, 171
Room and board, minimum wage and, 195

S

Safe harbor protection, 80–-97
  1099 filing requirement, 81–83, 272
  audit appeals and, 107
  basics, 42, 73, 80–81
  consistent treatment requirement, 81,
    83–87, 88, 105, 113
  for corporate officers, 75
  failure to qualify for, 109–110
  for household workers, 185
  limitations on, 96–97
  Obamacare penalties and, 215
  for prospective workers, 265
  for retirement plans, 97
  substantially similar positions and,
    84–87, 88, 93–94
  for workers' compensation purposes,
    174–175
Safe harbor protection, reasonable basis for
  basics, 88–96
  catchall provisions, 95–96
  court decisions/IRS rulings as, 88,
    91–92
  industry practice as, 88–89
  non-audit actions, 88, 94
  past IRS audits as, 88, 92–94
  professional advice as, 88, 90–91
  reasons that do not qualify as, 96
  VCSP and, 115
Safety
  insurance for household workers,
    191–192
  liability for employee injuries, 7, 157–158
  liability for IC injuries, 12
  OSHA protections, 9, 229
  workplace practices that support, 230

Salespeople
  direct sellers, 40, 46–47, 211, 218–219
  insurance brokers/salespeople, 72, 78–79,
    148, 165, 336–337
  outside sellers, 218–219
  real estate salespeople/agents, 40, 47–48,
    148, 165, 211
  traveling or city salespeople, 72,
    79–80, 147
Seamen, FLSA exemption for, 219
Seamstresses, statutory employee rules,
  75–76, 77
Self-employment taxes, 103, 245
Sequence of work, common law test and, 27
Services as part of cost of product/service,
  172–173
Services offered to general public
  common law test and, 22, 50, 55
  relative-nature-of-the-work test, 171
Services outside of normal business, ABC
  test, 138, 140–141
Services performed personally, common
  law test and, 28, 50
Services to be performed, IC agreement
  provisions, 294–295
Severability, IC agreement provisions, 308
Shareholders, 211, 248–249
Short-term workers, 254
Sick leave, 6, 31
Signatures
  electronic and digital, 287, 318
  IC agreements, 287, 288–290, 317–318
Skill and experience requirements,
  common law test and, 31–32
Social Security Act, employee classification
  rules, 9

Social Security Administration, misclassification audits by, 9

Social Security numbers, 96, 317

Social Security taxes. *See* FICA (Social Security and Medicare taxes)

Sole proprietorships, 244–245
  construction industry rules, 145
  IC agreement signatures, 288
  IRS audits of, 104
  workers' compensation and, 164, 177

Sports officials (amateur), workers' compensation for, 165

Spouses, payroll taxes for one spouse employed by another, 199–200
  See also Family members as workers

State agencies
  labor departments, 228
  for unemployment insurance, 135, 150
  See also Audits by states

State laws
  antidiscrimination laws, 228
  common law test reliance, 95
  copyrights for commissioned works, 237
  labor laws, 215
  new-hire reporting requirements for ICs, 280
  overtime compensation, 195
  real estate license requirements, 47
  researching, 189, 342
  which UC law applies, 135–136
  workers' compensation requirements, 160–161, 162–163, 179–181

State taxes. *See* Disability insurance/taxes, state; Income tax withholding, state; Payroll taxes, state; Unemployment compensation, state

Statute of frauds, 10, 284

Statutory employee rules, 72–80
  audits and, 106
  avoiding, 74
  basics, 42, 72–74
  for corporate officers, 72, 74–75, 147
  for drivers, 72, 77–78, 147
  for home workers, 72, 75–77
  for life insurance salespeople, 72, 78–79
  safe harbor protection, 73, 75
  for traveling or city salespeople, 79–80, 147
  unemployment coverage and, 147–148

Statutory independent contractors (statutory nonemployees), 40, 46–48

Stock options, worker lawsuits to contest, 128

Students, unemployment coverage for, 149

Substantially similar positions, safe harbor protection for, 84–87, 88

Switchboard operators, FLSA exemption for, 219

Systems analysts, safe harbor protections, 96–97

T

Tax court appeals, 108–109

Tax deductions, for retirement plans, 127–128

Taxes. *See* Backup withholding; Income tax withholding, federal; Income tax withholding, state; Payroll taxes, federal; Payroll taxes, state; Self-employment taxes

Taxi drivers, 68, 219

Tax laws
  federal, 337–342
  state, 342

Taxpayer identification number (TIN)
matching service, 266
Tax professionals, 336
  issues that require, 45, 335
  payment methods, 23
Tax reporting
  for corporations, 252
  for LLCs, 247
  for partnerships, 246
  for sole proprietorships, 245
  See also specific IRS forms
Technical service firms, safe harbor
  protections, 96–97
Telecommuting, 254
Temporary workers, 13, 149
Termination of IC agreements, 12, 307
Term of IC agreements, 294
Terms of payment. *See* Payment methods
Tools and materials, common law test and,
  22–23, 168, 170
  See also Equipment and facilities
Trade associations, business or
  industry, 337
Trade secrets, 234, 241–242, 324
Training, common law test and, 27–28,
  50, 53, 65
Travel and entertainment expenses
  common law test and, 25, 50, 70
  reimbursing ICs for, 269–271
Traveling or city salespeople, statutory
  employee rules for, 72, 79–80, 147
Treasury Department (U.S.),
  misclassification audits by, 9
Treasury Regulations, IRS, 340
Truck drivers, classification rulings, 69–71
Trust fund recovery penalty, 123–124

U

UC. *See* Unemployment compensation,
  state
Undocumented workers
  household workers as, 197
  illegality of hiring, 6, 197, 230
  immigration laws, 230
  workers' compensation for, 164–165
Unemployment compensation, federal, 4
Unemployment compensation (UC), state,
  130–150
  for agricultural workers, 144
  audits related to, 10, 103, 131–132
  basics, 1, 5, 130
  common law test and, 18
  for construction industry, 145–147
  for domestic workers, 144
  exempt occupations, 148–150
  for family members as workers, 201
  fighting claims for, 151–152
  IC agreement provisions, 305–306
  leased workers and, 253, 255, 259
  requirements for prospective workers, 264
  researching laws, 342
  for statutory employees, 147–148
  See also Classification for unemployment
  compensation
Unemployment tax, federal. *See* FUTA
  (federal unemployment tax)
Uniforms, wearing, 53, 65
Uninsured Employers Benefits Trust
  Fund, 158
Union rights, 6, 8
United States Citizenship and
  Immigration Services (USCIS), 197, 230
U.S. Department of Health and Human
  Services (HHS), 9, 206

U.S. Department of Labor
  DOL/IRS Pension Benefit Guaranty
  Corporation, 9
  FLSA enforcement, 192–195, 216
  FMLA enforcement, 224
  misclassification audits by, 8–9
  Misclassification Initiative, 216
  OSHA enforcement, 229
  unemployment insurance audits, 10
  website, 220, 342
  worker classification resources, 342
U.S. Department of Treasury,
  misclassification audits by, 9
U.S. Patent and Trademark Office, 233

V

Vacation, paid, 6, 31
Van operators, classification rulings, 67
Voluntary Classification Settlement
  Program (VCSP), 113–115
Volunteer workers
  FLSA exemption for, 219
  as statutory independent contractors, 211
  workers' compensation for, 165

W

Wage and hour laws
  Family and Medical Leave Act, 9,
  223–224
  National Labor Relations Act, 9,
  222–223
  overtime compensation, 192–193,
  195–196, 216, 217–219
  See also Fair Labor Standards Act
  (FLSA); Minimum wage

Washington
  unemployment coverage rules, 147
  workers' compensation classification,
  175–176
Websites
  EEOC, 220
  Fishing Tax Center (IRS), 48
  IRS, 212, 338, 343
  Nolo, 220, 342
  Obamacare, 209–210
  of prospective workers, 263
  state labor departments, 228
  state tax offices, 153
  state unemployment insurance
  agencies, 150
  USCIS, 230
  U.S. Department of Labor, 220, 342
Wisconsin
  UC classification test, 143
  workers' compensation classification, 176
Withholding. *See* Income tax withholding,
  federal; Income tax withholding, state
Worker Adjustment and Retraining
  Notification Act, 219
Worker benefits. *See* Employee benefits
Worker Classification Checklist, 264,
  355–357
Workers' compensation insurance, 158–182
  "assigned risk pools," 181–182
  audits related to, 10, 130, 132, 168,
  179, 248
  basics, 1, 5, 130, 156–157
  buying, 159, 181–182, 336–337
  for casual workers, 161–162
  common law test and, 18
  for construction workers, 176, 181
  for domestic workers, 162–163

exclusions for ICs, 157–158

exempt occupations, 160–165

for farm labor, 164

filing claims for, 12

IC agreement provisions, 304–305

"if any" policies, 178

for independent contractors, 178–181

leased workers and, 253, 255, 256–257, 259

relative-nature-of-the-work test, 170–173

researching laws, 342

for sole proprietors, partners, LLCs, and corporate officers, 164

state requirements, 160–161, 162–163

for undocumented workers, 164–165

who must be covered, 160

See also Classification for workers' compensation

Worker status. *See* Classification, IRS rules; Classification tests; Independent contractor status

Working hours

ACA employer mandate and, 210

common law test and, 29, 57, 65, 67, 70

full-time for one client, 57

overtime compensation, 192–193, 195–196, 216, 217–219

relative-nature-of-the-work test, 171

statutory independent contractors, 46

See also Part-time workers

Work-made-for-hire agreements, 322–323

Works made for hire, 235–238

by employees, 236

ownership rights, 239–240

specially commission by ICs, 236–237

See also Copyrights

Works of authorship. *See* Copyrights

Wrongful termination suits, 7, 12, 168

Wyoming, UC classification test, 143–144

⚖ NOLO *Online Legal Forms*

Nolo offers a large library of legal solutions and forms, created by Nolo's in-house legal staff. These reliable documents can be prepared in minutes.

Create a Document

- **Incorporation.** Incorporate your business in any state.
- **LLC Formations.** Gain asset protection and pass-through tax status in any state.
- **Wills.** Nolo has helped people make over 2 million wills. Is it time to make or revise yours?
- **Living Trust (avoid probate).** Plan now to save your family the cost, delays, and hassle of probate.
- **Trademark.** Protect the name of your business or product.
- **Provisional Patent.** Preserve your rights under patent law and claim "patent pending" status.

Download a Legal Form

Nolo.com has hundreds of top quality legal forms available for download—bills of sale, promissory notes, nondisclosure agreements, LLC operating agreements, corporate minutes, commercial lease and sublease, motor vehicle bill of sale, consignment agreements and many, many more.

Review Your Documents

Many lawyers in Nolo's consumer-friendly lawyer directory will review Nolo documents for a very reasonable fee. Check their detailed profiles at **Nolo.com/lawyers**.

⚖ NOLO Save 15% *off your next order*

Register your Nolo purchase, and we'll send you a
coupon for 15% off your next Nolo.com order!

Nolo.com/customer-support/productregistration

On Nolo.com you'll also find:

Books & Software

Nolo publishes hundreds of great books and software programs for consumers and
business owners. Order a copy, or download an ebook version instantly, at Nolo.com.

Online Legal Documents

You can quickly and easily make a will or living trust, form an LLC or corporation, apply
for a trademark or provisional patent, or make hundreds of other forms—online.

Free Legal Information

Thousands of articles answer common questions about everyday legal issues
including wills, bankruptcy, small business formation, divorce, patents,
employment, and much more.

Plain-English Legal Dictionary

Stumped by jargon? Look it up in America's most up-to-date source for
definitions of legal terms, free at nolo.com.

Lawyer Directory

Nolo's consumer-friendly lawyer directory provides in-depth profiles of lawyers all
over America. You'll find all the information you need to choose the right lawyer.

HICI8